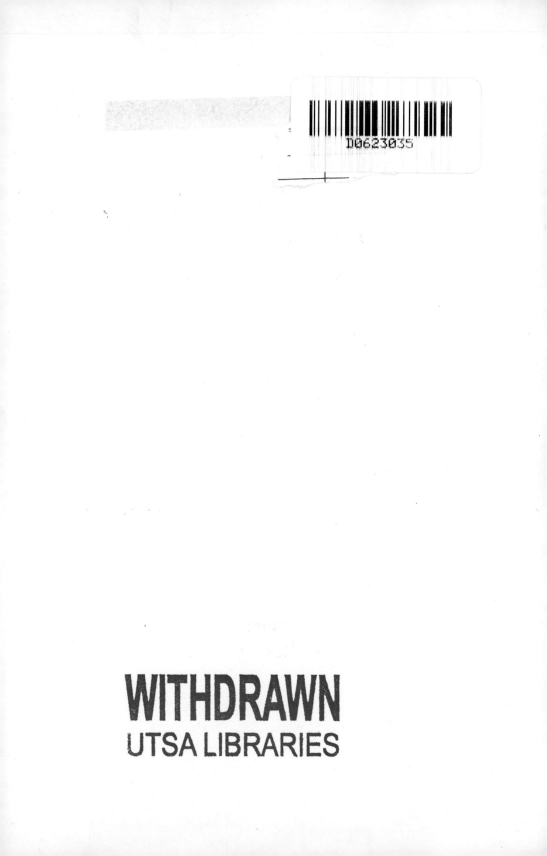

D0623035

Incentives for Environmental Protection

MIT Press Series on the Regulation of Economic Activity

General Editor
Richard Schmalensee, MIT Sloan School of Management

1 *Freight Transport Regulation*, Ann F. Friedlaender and Richard H. Spady, 1981

2 *The SEC and the Public Interest*, Susan M. Phillips and J. Richard Zecher, 1981

3 *The Economics and Politics of Oil Price Regulation,* Joseph P. Kalt, 1981

4 *Studies in Public Regulation*, Gary Fromm, editor, 1981

5 *Incentives for Environmental Protection*, Thomas C. Schelling, editor, 1983

Incentives for Environmental Protection

edited by Thomas C. Schelling

The MIT Press
Cambridge, Massachusetts
London, England

This book was set in Times Roman
by The MIT Press Computergraphics Department
and printed and bound by Halliday Lithograph
in the United States of America.

Library of Congress Cataloging in Publication Data

Main entry under title:

Incentives for environmental protection.

   (MIT Press series on the regulation of economic activity ; 5)
   Includes bibliographies and index.
   1. Environmental policy—Case studies.
2. Environmental protection—Economic aspects—Case studies. I. Schelling, Thomas C., 1921–
II. Series.
HC79.E5I513 1983   363.7   82–4631
ISBN 0–262–19213–6

Contents

Foreword

Government regulation of economic activity in the United States has grown dramatically in this century, radically transforming government-business relations. Economic regulation of prices and conditions of service was first applied to transportation and public utilities and was later extended to energy, health care, and other sectors. In the 1970s, explosive growth occurred in social regulation focusing on workplace safety, environmental preservation, consumer protection, and related goals. Though regulatory reform has occupied a prominent place on the agendas of recent administrations, and some important reforms have occurred in the energy and transportation sectors, the aims, methods, and results of many regulatory programs remain intensely controversial.

The purpose of the MIT Press series on Regulation of Economic Activity is to inform the ongoing debate on regulatory policy by making significant and relevant research available to both scholars and decision makers. Books in this series present new insights into individual agencies, programs, and regulated sectors, as well as the important economic, political, and administrative aspects of the regulatory process that cut across these boundaries.

This volume examines the use of economic incentives in programs aimed at environmental protection. The general case for such incentives is stated, three in-depth case studies are presented, and the attitudes of political decision makers toward the use of such incentives are surveyed and analyzed. Among the general conclusions that emerge are the importance of adapting the economic approach carefully to the detailed facts and problems of individual cases and the deep antipathy of many decision makers toward that basic approach. The three case studies make significant contributions to our understanding of the policy problems they address, and the volume as a whole provides valuable insights into the important but difficult task of employing economic analysis to design policies that will protect the environment in a sensible and efficient manner.

Richard Schmalensee

Preface

There is a discrepancy between the approach of economists to environmental protection and the approach of nearly everybody else. Activities that impinge on the environment are regarded by economists as activities that are, but need not be, external to the economic system—outside the market, unpriced, unowned, unmeasured, and not accounted for by those whose activity does the impinging. The impact is an economic "output" that differs from other outputs in not being comprised within private transactions. Economists recognize that there are problems in bringing these activities within the realm of economic accountability, but usually expect the problems to be solvable and do not see why the activities should not be subject to economic principles.

Economists treat good and bad environmental consequences as things that, in the historical development of property rights and obligations, have not yet been assimilated to that development or have been left to awkward bilateral negotiation when they might have been assimilated to the price system. Though economists acknowledge that many environmental impacts cannot best be managed through the price system, pricing is their first choice among management techniques; prohibition and other modes of regulation are exceptions to a general presumption.

But legislators, administrators, interest groups, and the general public perceive environmental impacts as arising from activities that are beyond the reach of business or consumer responsibility and in need of direct intervention. They do not perceive them as "outputs" in need of capture by that market system that mediates most of our producing and consuming. Indeed the market is blamed in the first place for motivating the behavior that needs to be brought under control. If the government is to intervene, the techniques of intervention that come to mind are not those that subtly moderate activities by manipulating perceived costs. And, as most of the impacts that concern people are adverse ones (the word *protection* testifies to that), pricing schemes look like schemes to compromise principles.

In politics, economists propose but noneconomists decide, and prices—or "incentives" generally—are not characteristic of governmental intervention. Whether the "environment" is air, water, scenery, and noise, or foods, medicines, and mechanical hazards, or the safety of the streets, or the racial composition of the schools, or speed limits, or fire regulations, or national parks, or the electromagnetic spectrum, most intervention is not designed to "tilt" decisions by some optimal manipulation of the relative costs of alternative behaviors.

Even economists would not insist that special license plates exempt from speed limits be sold by the motor vehicle registries, or that auto horns be metered so that drivers pay for the noise they make. Nevertheless, they are disappointed at how many opportunities for using the price system in environmental protection are overlooked or too readily dismissed. But most economists, including many who are conconcerned with environmental protection, are not veterans of legislatures or administrative agencies. They have not always done the empirical task of measuring what actual difference it might make to use prices instead of direct controls, and most of them have not devised in sufficient detail for implementation the schemes that their general reasoning leads them to admire and sometimes to advocate.

There are two polar possibilities here. One is that economists exaggerate grossly the virtues of the price system in environmental protection, underestimate egregiously the difficulties of implementation, and are bemused by their own theoretical constructs. The other is that they have failed to get their message across, or their audience is perversely or irrationally predisposed against their ideas, or there is some other removable impediment to the initiation of wise policies. The true state of affairs must be somewhere between.

This book was motivated by the following questions: Just how promising are pricing systems in environmental protection? Are they insufficiently appreciated, or are they less workable than they appear? What are the objections against pricing proposals? Whose are the objections? And are the objections sound or unsound? If some objections are unsound, what might be done to make better use of pricing and incentive schemes in areas where good use could indeed be made of them? The motive was neither to clear the way for successful promotion of incentive systems nor to deflate them once and for all. A discriminating analysis might discover the generic characteristics of those environmental problems for which pricing and other incentive schemes will work and will actually make a difference compared with alternative schemes and those in which pricing systems are idealized mechanisms that could not work or would not make enough difference to be worth the trouble compared with alternative modes of intervention.

Of the three case studies presented here, two concern local "emissions" that are candidates for control via incentive arrangements: benzene, a chemical known to be potentially hazardous to human health, and airport noise, a source of discomfort, reduced property values, possible impairment of hearing and health, and political conflict. These are two very different candidates for pricing measures of the kind generically known as "emission charges." The third case study relates to local air pollution from stationary sources and examines the prospects for "marketable rights" in emissions. Each case study was the full responsibility of the individual author, each is presented here as a self-contained treatment of its subject, and each examines a specific current policy issue and explores a common type of problem. The three constitute a small but diverse sample of environmental controls.

Reduced to a paragraph, here is what each of the studies finds:

• In 1977 the U.S. Environmental Protection Agency listed benzene as a "hazardous air pollutant," primarily on evidence that it is a leukemogen. Since then, the EPA has been developing emission standards for specific source categories. The first, for maleic anhydride plants, was proposed in April 1980. Analysis of costs and benefits suggests that the proposed standard would not be cost-effective and that alternatives, particularly a charge based on plant-specific exposures, would be more efficient. Such an approach could be extended to other major stationary sources, offering gains in efficiency and timeliness over source-by-source-category standards. Other sources, such as automobiles and service stations, also suggest opportunities for incentive-based schemes (primarily charges).

• Several characteristics of aircraft noise make the charge approach attractive. Unlike many air and water pollutants, noise permits the use of existing information to set charges that approximate marginal damages (and without the need to "put a value on human life"). An aircraft-noise charge also can be administered easily, because few sources are involved. Noise standards allow modifications (primarily changes in the timing of the requirements) to minimize any disadvantages. Especially important is modifying noise control to permit different regulations for different airports. The revenue raised may not be well spent, or welcomed, by the Federal Aviation Administration. The noise charge is a much less attractive option from the perspective of the airport operator for whom the noise charge's decentralization of decisions on the means of achieving noise reduction is a disadvantage. Indeed, though marketable rights are a less attractive option than the noise charge at the national level, the rights scheme (or a modified version) may be more attractive for an individual airport.

• The Clean Air Act limits further pollution in "clean air" areas, and assigns states responsibility to allocate rights to new sources to use this assimilative capacity. Case study 3 considers ways of assigning these rights: giving them away, auctioning them, creating markets among existing and new sources, and variants of these. The main finding is that the present method of giving rights away contributes little if anything to efficient regulation, but that the consequences of the other alternatives depend on the characteristics of new-source growth and existing emitters in the region where the policy is applied. It is impossible to rank alternative incentive systems without reference to a specific market setting.

The study of attitudes is based on interviews with members of the House and Senate staffs and staff members of Washington offices of environmentalist organizations and trade associations. The interviews showed little knowledge of efficiency arguments in favor of charges. Proponents of charges tended to endorse "the market" in a general and ideological way, deprecating government; opponents were uneasy with or hostile toward "the market" and favored government intervention. Implicit and explicit considerations of the political appeal of different formulations and approaches were evident.

The three cases are so different that the conclusions, insights, discoveries, and suggestions reached in one may not have counterparts in the others. Time of day, for example, is crucial for noise but not for benzene. Marketable permits for air-quality control depend on how "thin" the local market is, but market structure matters little for benzene. Control of benzene can depend on enforcement at tens of thousands of service stations and fewer than a dozen large plants, whereas most air pollution comes from a few large sources. Victims of aircraft noise know that they are victims; victims of benzene may have no idea what the damage is. And so on. Most of the interesting insights and conclusions are at least partly specific to noise, chemicals, or air quality, and though many of them can be extrapolated, they are best summarized in the individual studies.

Nevertheless, several themes or results do show up in all three studies or in two of the three. Since the cases were chosen for their differences, conclusions that are applicable to more than one are of special interest. They will not be elaborated in detail here. Some are judgments that the reader can share only through reading the case studies. Some require qualifications that are best formulated in the context of the studies. But they can be anticipated here. The purpose of stating them is to alert the reader to them, not to make them sound self-evident in advance.

That they are not self-evident is attested to by the fact that in the early stages of planning the authors did not foresee these common conclusions.

Deserving first mention is that, after careful examination, pricing mechanisms for environmental protection are found practical. That they were theoretically attractive in advance should not be surprising: the authors are economists. As mentioned above, it was the attractiveness of pricing to professional economists and its unattractiveness to noneconomists that motivated the project. A likely hypothesis was that economists were aware of all that is attractive about pricing mechanisms, and what is attractive comes out of economic theory, whereas administrators and legislators were aware of all the practical reasons why pricing would rarely work well enough for the theoretical virtues to outweigh the practical objections. But it does not appear from these three studies that pricing, whether by charges or by marketable rations, is impractical. Administrative difficulties did not prove to be serious objections to charges in contrast to standards. Although there are problems (some of them difficult), the main problems are common to pricing mechanisms and regulatory standards.

From the preceding observation comes a second result of the studies: The impediments do not differ as much as one might have thought between emission charges and emission standards. Uncertainties are often serious, but serious for both charges and standards. Monitoring of emissions can be costly, difficult, or unreliable, but monitoring aircraft noise in order to impose charges requires the same equipment as monitoring in order to enforce standards. When benzene emissions from maleic anhydride plants cannot be easily monitored regularly, standards may have to be applied instead to the control equipment. Then, however, charges can be assessed against what equipment is installed with the same reliability. A charge on gasoline stations can be based on the type of emission-control equipment installed as well as the amount of gasoline pumped. And if permits (marketable or nonmarketable) can be monitored and enforced, there is no added difficulty in offering them for sale. Ignorance of how damages relate to emissions, and disagreement about the severity of damages and the money value of reducing them, impose difficult or even insurmountable problems; but when the problems are insurmountable, they are for both charges and standards.

A third conclusion emerges that the reader may not have expected any more than the authors. It is that many of the benefits from using well-designed pricing mechanisms can be obtained with sensible, well-designed regulatory standards. Among economists there may be a tendency to compare existing regulatory standards with optimally designed charging systems. But what comes out of the first conclusion mentioned above is not that optimally designed, perfectly administered charge schemes are practical; it is that the compromises that have to be accepted to make a pricing mechanism workable do not compromise away all the advantages. Still, any pricing mechanism will be full of compromises. And it is fair to suppose that, if the principles underlying pricing were accepted by legislators and administrators, the same principles could be reflected in regulatory standards. So the case for pricing is not quite as strong as the first conclusion made it sound. Sensible standards based on the same criteria as pricing can often accomplish many of the same benefits. What the studies show is that there are usually some striking improvements to be had in costs and benefits by adopting the right criteria for standards.

This deserves a little amplification. A point that will recur throughout the book is that even in contemplating a pricing mechanism, and especially in designing one, we are ineluctably led to assess values. The values are costs and benefits. The benefits take the form of reduced harm and damage of some sort, and the damages have to be identified, measured, and related to whatever is going to be controlled, regulated, or charged for. Noise itself is not evil; it is what it does to people that is bad. How bad depends on how many people there are, how they live, how vulnerable they are, and what means they have of protecting themselves. Benzene can cause cancer, but how much cancer a given concentration in the atmosphere for a given length of time can cause depends on how many people are exposed and on the age, health, and other characteristics of the life expectancy of the exposed people. Proposals for charges compel attention to the damage, not to a quantity emitted or to some average atmospheric concentration. Proponents of regulatory standards, in contrast, seem not to be so strongly attracted to a measure of damage as the focus of concern. The quantity of emissions, or else the ensuing atmospheric concentration, is often taken as the ultimate target.

The most critical difference between focusing on a measure of damage and focusing on emissions or concentrations is likely to lie in the attention given to population density or total exposed population. It is hard to find any methodological or philosophical issue in environmental protection that is more divisive or more important than whether exposure ought to be minimized in the aggregate or equalized among individuals in different localities and situations. The issue can be avoided, or at least evaded, if costs can be neglected and all exposures of any significance are deemed unacceptable. But when costs cannot be neglected as insubstantial, or when at no feasible cost can all significanct exposure be eliminated, this inherently distributive qucstion has to be faced. And that it is unavoidable does not make it easy to answer (even for most economists).

Anothci difference is otten in the use in charge schemes of the cost and effectiveness of the alternative defensive measures (such as noise insulation) that communities or individuals may take, in attaching an upper limit to the worth of reducing an offensive substance or activity that can be defended against instead.

It is not universally accepted that these considerations are appropriate. When economists conclude that sensible regulatory standards based on the appropriate benefit and cost criteria can accomplish much of what charge schemes can accomplish, not everyone will agree with them on what are sensible and appropriate criteria. The concluding chapter of this book, a study of attitudes, will drive that point home. People who object to charges are likely to be also opposed to standards that derive from comparisons of assessed damages and abatement costs.

The next conclusion is that exposure and damage differ enormously from place to place. The harm that emissions of a particular kind can do can differ by a factor of 10, or sometimes 100, even for emissions from similar sources. The main difference is usually the population at risk. An obvious implication is that emission charges at some sources should properly be 10 percent, or only 1 percent, of charges on the same quantity of emission in another locality. Evidently this would mean that charges in some localities would be rounded to zero. The corollary is that sensible regulatory standards, too, not just charges, if properly oriented toward damage rather than emission quantity, should vary by location in the same way.

The point made above about the problems of practicality being shared by standards and charges shows up here. Charges that vary per unit of emission in order to be uniform on damage may run into political obstacles. For purpose of administration there may have to be some small number of exposure classifications for different sources rather than continuous variability according to current exposure data. But exactly the same can be said about varying emission standards.

A difference between charges and standards shows up in connection with varying costs of abatement from source to source. A given level of aggregate abatement will be achieved at lower aggregate abatement cost if the firms abate most that can do it most cheaply while the firms whose abatement is exceedingly costly may not even abate at all. (Large differences in abatement costs—differences by more than a factor of 2 or 3—are likely to arise among sources that, while emitting the same substance, are in different lines of business; for example, firms that produce a substance and firms that use the substance as a raw material.)

In a single locality where all emissions contribute the same to damage, a uniform emission charge will appear nondiscriminatory while allowing the firms least able to abate their emissions to save money by paying the charge and allowing the firms best able to abate to save money by reducing emissions. Marketable emission rights do the same, of course, if the aggregate quantity of emissions allowed is compatible with the price that would be charged in the alternative regime. It is hard to achieve that result with emission standards—excusing firms if they claim high abatement costs, and obliging those that abate successfully to abate even more. It can be done if the firms with the high and the low abatement costs are engaged in wholly different operations and can be classified separately for environmental purposes. Even then the standards will suffer from any uncertainty about abatement costs. Charges and marketable rights leave the abatement-cost calculation up to the individual firm. An airline could make its own calculations whether to retire or to retrofit or to reroute certain aircraft, or to postpone retirement and pay the charge.

The airport-noise and air-quality studies display another dimension on which differential charges on exposures at different locations can make a difference: Business may move to where exposure is less and charges are therefore lower. (With airport traffic, time-of-day variation is as significant as locational variation.) Air traffic can be rerouted in response to differential charges more readily than plants can be moved; however, new plants seeking air-quality emission permits may choose locations where they can purchase permits least expensively from other firms or from issuing agencies. Generally this should be seen as a favorable environmental result. Locating activities where they can do the least harm, because of the smallest exposed population, is a benefit in the aggregate to the total population. It is not a benefit, however, to the people who live where they haven't enough neighbors to provide a high-exposure classification, and therefore a high emission charge, to keep emitting firms out. This is the issue of distributional equity that was mentioned in connection with the focus on exposure and damage.

But something else often goes with it. When airline business moves to other cities, or firms build new plants elsewhere rather than enlarge existing plants, jobs and earnings are affected. There are the jobs and earnings immediately affected by the relocation and the further effects of reduced consumer spending in the localities from which people move or in which jobs are lost. In some cases the majority of the local residents might be happy to see a noxious plant close down, and only the most directly affected residents would be dislocated. Or perhaps the activity, noxious as it may be, contributes enough economic benefits so that nearly everybody would feel the loss that would result from cleaning up the environment at the expense of a major source of income.

Nationally, the shifting employment may all cancel out; one city's loss of jobs is another's gain. But local interest in differential charges or standards, with their attendant shifts of business and jobs, can be intense. High charges (based on high population density around an airport) that threaten to reduce airport business may be unpopular, even with those who live near the airport and its noise. Regulatory standards may be more popular, but only if they are uniform, paying no attention to damage. Thus, standards may be more acceptable when they provide less protection against environmentally mediated damage and more protection against economic damage. "Sensible standards" would not have that acceptability advantage. Still, an issue arises that is not easy to dispose of when the population with the most to gain from removal of some noxious activity also has the most to lose, and would choose not to lose it.

Another conclusion, one that was somewhat anticipated but emerged more vividly than we expected, is that the revenues produced by any charge scheme can be an embarrassment. (This is separate from the point that emission charges in the aggregate may or may not be more costly to the affected firms than standards that achieve the same percentage reduction in emissions.) In principle the proceeds ought to be a good thing. They come out of a scheme that promotes efficient economic behavior and could be used, say, to replace taxes that had deleterious incentive effects. But that depends on politics, jurisdiction, legal arrangements, and budgetary behavior. An agency that sells emission permits may never have been given the jurisdiction or authority to make good use of the funds. And what treasury it delivers them to may be determined by legal arrangements made before there was any intention that the agency have large amounts of money to disperse or surrender. What happens to the proceeds from aircraft-noise charges may well depend on the level of government at which the charges are imposed—local, state, or federal. The opportunities for misspending funds cannot be disregarded in reaching a judgment on any particular scheme for noise charges.

Marketable permits limited as to total quantity, pertinent in the case of aircraft noise as well as that of air quality, make sense primarily when damages increase more than linearly with the damaging substance or activity. However, for air quality and aircraft noise this nonlinearity is always regional or local. A national market is economically attractive and would usually obviate the dangers of market thinness or manipulation, but the purpose of a market in emission rights will almost always be to create a local "bubble" (like schemes that permit multiplant firms to "consolidate" the emissions of several plants and meet a standard for their collective output). So trading has to be confined within each region or locality. But if each locality or region has to constitute a separate and independent market for emission permits, the markets may be too thin to support competitive bidding. The thinness of the market depends largely on the number of firms, and somewhat on whether firms are commensurate in size or identified with the same industry or process. Thus, the way that permit schemes work will often depend on the structure of the product market. This is a strikingly distinguishing characteristic of marketable permits in contrast to emission charges.

Thus, besides features common to two or all three of the cases, there are contrasts that are equally instructive. There are, for example, major differences in the breadth of potential responses to the environmental problem. Aircraft noise offers a wide range: reducing the noise, removing the noise, removing the victim, insulating the victim, insulating the engines, and arranging for the noise to occur when the victim is absent are altogether different in what they do but not in what they accomplish. And pure financial compensation to the victim is likely to be less objectionable if the damage is nonlethal and if it is distinctive, identifiable, immediate, and universal, as with aircraft noise, rather than remote, statistical, and small compared with background damage, as with airborne benzene.

These, then, are some of the commonalities and contrasts to watch for when reading the case studies. And I should stress that case studies is what they are meant to be. Although each is a comprehensive examination of a particular environmental issue, each alone and the three together are intended to reveal principles and ideas that are of some universality. My editorial judgments about the scope of the studies, the depth of detail, the state of the art in monitoring emissions or assessing damage, and the administrative judgment or political compromise called for reflect my intention that these studies, singly if necessary but preferably together, should illuminate principles and problems of some universality.

The introductory chapter lays out what might be called the economists' case in favor of treating environmental ills as susceptible to economic diagnosis and, in at least some cases, to resolution through a pricing mechanism. Some effort has gone into trying to make this approach familiar and plausible rather than esoteric. There are familiar "market solutions" to some environmental problems—or to what used to be environmental problems before solutions were found that we may now take for granted and neglect to recognize as solutions to environmental problems. An entire genus of environmental problems and their possible solutions is epitomized by the lowly and ubiquitous parking meter.

Chapter 1

Prices as Regulatory Instruments
Thomas C. Schelling

Someone who spoils his own land without affecting drainage on other property, runs noisy equipment that no one else can hear, or contaminates his own water supply is not said to create an environmental problem. The problem is said to be environmental when lead and sulfur drift downwind to make somebody sick, an oil spill washes onto a public beach, acid drainage from an abandoned mine destroys marine life, or the burning of fuels changes the regional or global climate. Environmental effects are the consequences that are outside the purview, the cost accounting, the concern, or the responsibility of identifiable producers and consumers. They are outside the pricing system (except when damage suits can make the activities costly).

The activities that have environmental impacts are of many kinds. There are beneficial activities that go unrewarded and harmful ones for which neither price nor penalty is paid. There are some so natural, like the happy voices of children a little too early in the morning, that we hesitate to complain. However, the roar of a motorcycle gets little sympathy. There are individuals who pour crankcase oil down the storm sewer or throw cigarette butts out the car window, and there are businesses, large and small, that dump their waste on land and sea and in the air. There are activities, such as driving and parking, that are unobjectionable on a modest scale but on a larger scale cause obstructions. There is illicit disposal of hazardous waste, disposal that is legal but dangerous, and unwitting disposal of things that prove harmful. There are activities, such as burning fuel to make electricity, that are bound to continue, the issue being only the scale of the activity. Other activities, such as the use of certain herbicides or pesticides, need not and ought not to continue at all. There are damaging activities that are so beneficial that the harm is incidental and no "problem" is perceived, activities that pose difficult choices because of disagreement or uncertainty whether the benefits or the damages are greater, and activities that almost without question deserve to be discouraged or stopped.

And when all those with any conceivable stake in the matter or any right to participate get together and determine that some investment in the environment is not worth the cost or that some depletion is more than compensated by the benefits, they may count the quorum unmindful of all the interested parties who have not been born yet.

Most of us play many roles on the environmental stage. We are assaulted by construction and aircraft noise, offended by roadside advertisements, endangered by lead and mercury, and delayed by double-parked vehicles. Our children are threatened by broken glass when they go barefoot. But somebody else is offended by our children's bare feet, and if that broken bottle wasn't ours we may have seen it and neglected to remove it before it got broken. We ourselves contribute to the Christmas-shopping congestion. Our chamber music annoys the teenager as much as her disco music bothers us. We rarely ask our neighbors what color they would like us to paint our houses. Some of us smoke and inflict carbon monoxide on others in an elevator; some of us quit and chew gum, which may smell as bad as cigarette smoke. And if our automobile is fully insured against theft we may carelessly let it be stolen and abandoned or burned in a vacant lot. We complain about the high price of electricity while the utility companies try to keep costs down by avoiding expensive smoke abatement. We patronize an economical laundry that pours chemicals down the sewer. We pay higher prices for oysters without knowing that, whether or not the oystermen are suffering from pollution, the oysters are and it makes them expensive.

It takes some analysis to identify victims and beneficiaries. When electric utility companies are required for the sake of health, comfort, and agriculture downwind to clean up the hot gas that comes out of the smokestacks, it is not easy to figure out who is going to pay for those downwind benefits. A good guess is that, taking all the coal-burning electric utilities together, it will not be their stockholders. Electric rates are set by public boards and commissions that determine what costs are deductible and what rate of return is appropriate, and most of the cost is passed along to customers. Some of the customers will be industries; they, too, will pass most of the rate increases along to their customers. If the industries use much electricity, business may suffer some and wage increases will be less forthcoming or property-tax assessments less aggressive; and all the results are as complex as locating the ultimate incidence of a tax on electricity or on the coal used in utilities' boilers.

These complications remind us that efforts to preserve or improve the environment, protect against hazards, and abate the degradation of soil and forests and other resources entail a mix of gains and losses and of gainers and losers. And neither the aggregate size of gains and losses nor their distribution among gainers and losers is transparent. The distributional issues—who gains and who loses, by income size or region or occupation—are not easy to discern and are sometimes contrary to intuition. Even the relative magnitudes of gains and losses (those that accrue to individuals and those that are thought of as social values) are not easy to estimate. Not only is the analyst or observer hard put to be sure who is affected by how much in which direction, but even the victim of environmental damage (and the beneficiary of protective measures), like the beneficiary of environmental permissiveness, can be unaware of the magnitude of his gain or loss—even unaware of any concern in the matter. And those on whom the costs of environmental protection initially fall can have an exaggerated sense of being the ultimate victim of environmental policy, not recognizing how many of the costs get passed along to customers or suppliers.

Although this kind of analysis is difficult in practice, the theory with which to think about externalities is well developed. And in theoretical terms it matters little whether the externality in question is one that we would call environmental. Nevertheless, in assessing those impingements of people's activities on the welfare of others that are mediated by the physical environment, there are some elements that the theory of externalities cannot handle. The reason is that the theory is neutral toward the moral significance of activities and events, and excludes values that cannot ultimately be assessed in individual welfare.

Specifically, economic theory evaluates actions by their consequences and by the way the consequences are valued by the people who benefit or suffer, not by whether they are inherently good or evil or according to the spirit in which they are done. However, for many people concerned about the environment, activities are not adequately measured by a summation of individual gains and losses. That there is no one to speak for a particular endangered species or for the Earth itself does not, for some people, imply that, because nobody has a stake in the matter, there is no matter. And certain actions that can do grievous injury to innocent people or other living things are not always assessed (and possibly dismissed) according to the amounts of harm that accrue to individuals. An offense may be unpardonable independent of its consequences.

Environmental concerns are loaded with additional weight because they occasionally, as with climate, pose the possible risk of irreversible damage of an immensity difficult to agree on. And they entail chemical and biological hazards capable of causing, on a large or a small scale, those most awesome human tragedies, death and fetal abnormality.

Externalities and Social Controls

So it is that people and businesses engage in activities that use up something scarce that doesn't belong to them, spoil something in which others have a legitimate interest, or harm other people directly. Or they do these things immoderately when on a modest scale they might be excused. And they can do it because they do not know they are doing it, or they know but do not care, or they know and care but have a conflicting interest of their own, or they do not know how to moderate their activities or redirect them in less harmful ways. They may enjoy the anonymity that allows them to get away with it, or they may regret their inability to identify their victims and make amends. And activities beneficial to the environment may go unrewarded for similar reasons.

Thus arises the need for some social control, some inducement to curtail or moderate the harmful activities or to enhance the beneficial ones. The generic term *environmental protection* implies that it is the harmful activities that deserve most urgent attention.

Self-controls

For want of a better term we refer to the various ways environmental policy is implemented as *social controls*. The term has to include what might better be called *self-controls*, which are much relied on in social policy. It must include education that generates awareness and knowledge of the consequences of actions or the availability of alternatives. It includes prohibitions, regulations, fees, charges, tax incentives, and methods of enforcement. And it includes what may be done through private negotiation, civil law, and informal cooperative arrangements.

Unquestionably a large part of environmental protection depends on decency, considerateness, and self-control. Some of the most powerful social controls work through habit and etiquette. Three generations ago spitting was common, even indoors, and sanitary protection took the form of strategically placed containers; today, many elevator entrances are guarded by containers of white sand for the receipt of burning cigarettes. Two generations ago the spittoon had disappeared, but "No spitting" signs were everywhere, even on streetcars, and violations of the no-spitting ordinance were threatened by fines the way littering from a car window is today. One generation ago spitting had ceased to be a menace, probably because the technology of nicotine ingestion had shifted from plug to cigarette. Then butts became the problem and ashtrays the solution. Today the emphasis has shifted from the debris of smoking to the smoke itself. We have gone from "No spitting" to "No smoking" in half a century, but except in airliners and some theater lobbies enforcement is more by moral suasion that regulation. The same was true of spitting 50 years ago, though, and today it is not the fear of that $50 fine that keeps people from throwing litter out the car window.

Considerateness and self-control deserve emphasis. A propensity for civilized behavior is a precious resource for protecting our environment, and the more overt social controls should be compatible with preserving that widespread sense of personal responsibility on which, especially in densely populated areas, we all depend so much.

Overt controls
The more overt controls take a variety of forms. There are prohibitions on the transport of dangerous chemicals and regulations on where campfires may be lit, when noise may be made, how wastes may be disposed of, how much gas may be emitted, how many deer may be killed, what ingredients may be used in a pesticide, how fast a car may travel, how loud a noise may be, and what unsightly or dangerous objects may not be left in a public place. These can be enforced by fines, imprisonment, tax penalties, or loss of licenses and privileges.

Then there are the fees and charges collected at parking meters, toll booths, and entrances to national parks. There are tax credits for saving energy, weight-related license fees on heavy trucks, federal subsidies for municipal waste disposal, and even bounties for vermin and old containers picked up and delivered. (In some places there are finders' fees for leftover explosives from earlier wars.) And there is the civil justice system that allows private citizens to bring suit against those whose environmentally offending activities, licit or illicit, violate their rights and inflict harm. And the ballot box is a medium of social control when government is the offending party.

There are many ways to classify these media of control. They can be formal and legal or informal and voluntary, rigid or flexible, enforced or merely promoted, based on negative or positive sanctions, aimed at activities or substances or damages, universal or specialized, and with different degrees of reliance on citizen enforcement or enforcement by police power.

This chapter is about a particular set of measures of social control — measures of a kind familiar in some contexts and unfamiliar, even alien, in others. They are sometimes described as economic incentives, sometimes as measures that work through the price system. A list of measures of this kind would include fees and charges, subsidies, rewards, indemnities, auctions, the assignment of property rights, and the creation of markets. Familiar examples are parking meters and bottle deposits, income-tax credits for home insulation, stimulated markets in recycled paper, and low-interest loans on solar home-heating systems. Less familiar and more controversial are fees, fines, or taxes on the sulfur emissions of electric utilities, on airport noise, and on the venting of noxious fumes. Attempts to generate markets in which new firms that cannot operate without emitting into the atmosphere some quantities of regulated substances can buy the rights of existing and former firms, which in turn discontinue equivalent emissions, are a more recent innovation, as is the proposed "carcass deposit" on new automobiles, which would be refunded at the end of a car's life to whoever disposed of the remains at an approved site.

Pricing systems
The distinguishing characteristic of these measures is not just the payment of money in consideration of damage done, resources depleted, or environmental benefits conferred. I exclude from this study the widely advertised fines for mutilating subway seats, throwing trash out a car window, or letting one's pet do what is known in London as "fouling the footpath." Fines levied on or damage payments collected from companies that carelessly or surreptitiously dump hazardous substances illicitly in rivers and ponds and vacant lots are not what I mean by prices, even though the anticipation of a fine may add an economic incentive for staying within the law. And rewards paid for information leading to the arrest and conviction of whoever set the woods on fire, defaced public property, or dumped toxic waste by the roadside at night, however much economic incentive is intended in getting people to do their clear public duty, are not what I would call price mechanisms.

The distinction is illustrated by the difference between a *fee* and a *fine*, or between a *charge* and a *penalty*. It is typical of fees and charges (whether a fee is collected on an activity or paid for its performance) that no moral or legal prejudice attaches to the fee itself or the action on which or for which it is paid. The behavior is discretionary. The fee offers an option: Pay it and deposit your rubbish here or take your rubbish elsewhere; pay it and park here or drive someplace else; pay it and swim at the public beach or don't swim. Whether used to cover the cost of cleaning up afterward, to maintain a public facility, to keep the facility from being overcrowded, or to indemnify those who may be discomfited or whose productivity is adversely affected, a fee entitles one to what one has paid for. It is not levied in anger, it does not tarnish one's record, and even if paid to cover damages it is not expected to be paid grudgingly and received resentfully. But a fine that is paid upon conviction for an offense does not erase the offense. "Paying one's debt to society" is not an apt metaphor. There is no thought that society breaks even when someone serves a prison term or pays a fine. The conviction stays on the books; the law has been violated and the behavior is reprehended.

The distinction is clear in principle, even though fines for traffic violations are modest in amount, excused or reduced for the first few offenses, and not counted as a police record. There may be cases in which willingness to pay a fine is a test of personal urgency, and the violation is publicly condoned or legally excused. But even then the action usually has to be excusable on its merits; the paying of the fine is only an earnest of seriousness and not a license for misbehavior.

Decentralized decisions
The essence of a pricing system is that it leaves the decision to pay or not to pay to whoever confronts the price. The price is there to be paid—not to stand as a warning, an act of redemption, or an admission of liability. A pricing mechanism decentralizes the decisions. A central authority may determine the price, but paying or not paying is a decentralized choice. In contrast, a central authority determines that you may not park in front a hydrant—not you, not anybody, not for an hour, not for five minutes, not to make a phone call, not to go to the bathroom. The fine is an enforcement device. It may be calculated to deter; it may be thought commensurate with the offense; but there is no intention that, with the fine for an offense set, it is now freely up to you whether or not to park there.

In a similar way, some taxes on damaging behaviors are excluded from the category of price mechanisms. A tax intended to reduce some consumption or activity may embody disapproval and be equivalent to an imperfectly enforced prohibition. A tax is described as punitive when its political motivation makes it akin to a fine. (Taxes on liquor and tobacco reflect disapproval.) There are other taxes that are reminiscent of fees and charges because they are related to wear and tear or damage generated by the taxed activity, or to the cost of providing facilities for the activity. Gasoline taxes earmarked for public roads are like indirect fees to maintain highways.

What is crucial about a price is not merely that it is devoid of ethical significance or that it allows freedom of choice. A price is a *measure*. In the marketplace a price on a voluntary transaction must cover the cost of performance or provision, or represent adequate compensation. The transaction is voluntary. The price paid for a bottle of cola must cover the cost of labor, materials, energy, and anything else that went into producing it, delivering it, and handling it up to the time of sale. If people will not pay enough to cover those costs, we conclude that the drink is not worth what it costs; the labor, materials, and energy will be used elsewhere to produce things that consumers do consider worth the cost. And if the cost of cola delivered to the neighborhood store is well within the prices that consumers will pay, and if no single firm has a monopoly, we expect competition to produce all the cola that people want at a price that covers its cost. Thus, the amount of cola produced and sold will be the amount that consumers find worth the cost, where the cost reflects the alternative uses of the resources that go into making and bottling and distributing cola. The price serves simultaneously as a measure of the worth of the resources that go into the cola and the worth of the cola to consumers (more accurately, the worth when *that* amount of cola is being consumed — smaller quantities are valued more highly, whereas additional quantities would be worth less than the cost; otherwise people would be buying more).

Sometimes prices reflect a scarcity value. Most cities regulate the number of taxicabs. Under the medallion system there is some fixed number of licenses. Medallions are salable, like seats on the stock exchange, and the question of who drives the taxis is thus decentralized to the marketplace. There is a going price in each city. A medallion must be worth at least that price to its owner (because he has the option of selling), and must be worth less than that to anyone else (because otherwise somebody would bid the price up until the number of people to whom medallions were worth that much was the same as the number of medallions).

Thus, what is decentralized when goods and services and privileges are "priced" is decisions about who consumes or produces a service, which services are produced and consumed, where or when, and how much. Actually, nobody ever decides how much cola should be produced or consumed; each individual decides only how much he or she will produce or consume. For taxi medallions there is a centralized decision on how many, but the market determines who.

To translate the examples of cola and medallions into the environmental area, we can think of "bottle bills" requiring a deposit to be paid on returnable containers for beer and soda. Two generations ago beer and soda were sold in returnable containers. Most of us gave little thought to whether, having paid the deposit, we owned the bottles or had merely rented them and pledged with a small deposit to return them. No matter. We paid enough deposit to cover a replacement if we lost or broke or kept the bottle. There were children who would collect bottles for profit if we thought them not worth transporting back to the store. And if we lost or broke one we never had to say we were sorry; the price covered the cost. Neither the store nor the manufacturer cared what we did with the bottle.

But somebody did care, or does now. Broken, the bottle was a nuisance, and intact it was unsightly. If I left a broken bottle on your sidewalk, I had paid for the bottle, but not for cleaning your sidewalk, or repairing your bicycle tire, or treating your dog's paw. But because the bottles cost money I was careful, and if I left behind an unbroken bottle some youngster might retrieve it and redeem it at the store for two cents before it got broken. Now that our standard of living is so high and the technology of disposable containers so cheap that most of us prefer throwaways, the failure of the price of the bottle to reflect the costs of cleaning up or the compensation for damage done is a greater discrepancy than it used to be. The deposit on a returnable container has been reinvented for regulatory use.

Actually the deposit, as a control mechanism, need not be associated with reuseable containers or even recyclable materials. And a deposit need not even be money; one could leave one's name, address, and Social Security number and be subject to a fine for failure to return the bottle or an equivalent amount of broken glass. The deposit system is just an artful convenience. It is a system of decentralized automatic enforcement. It imposes a penalty that is devoid of ethical significance for any bottles not returned. And because bottles may conveniently be returned to stores that handle bottled goods, and the same trucks that bring fresh drinks can carry empties back where they came from, reuse is an economical byproduct of the enforcement scheme.

Price determination   But how much should the deposit be? At a minimum it should cover the cost of the container, of course; else the consumer is not even paying for the contents. And if, having paid a deposit that covers the replacement cost, consumers husband their bottles and return them faithfully, the problem is solved. If instead they leave them in large numbers at the beach and on the sidewalk, we want a larger deposit to induce a higher rate of return. If we collect a nickel a bottle and it makes no difference, we know that the nuisance of returning empties is equivalent to more than a nickel a bottle for most consumers. Somebody is getting revenue without much being accomplished. We could raise the deposit to $1 a bottle and the return rate might reach 100 percent. The system does not cost people anything in money outlays, because, the cost of wasting bottles being so high, they are not wasting them and paying any money. They are losing only the interest on a few dollars' investment. But we have no measure of the trouble people are going to to preserve their bottles; $1 merely sets an upper limit. (And if we were hoping to get a little money into a cleanup kitty from the unclaimed deposits, we are not getting it.) If $1 seems high enough to be punitive, we can think of it as a penalty rather than a price.

We could even make the deposit $5 per bottle, so that a six-pack would cost as much as a major traffic violation. At that point the deposit would clearly not be a price. It might still be a splendid method of enforcing a ban on littering. Nobody would have to make a citizen's arrest of a litterer, bring the discarded bottle for evidence, and take up a court's time for the imposition of a fine; it would all be handled automatically by the deposit, as long as we didn't insist on criminal records. But at $5 we would surely not be setting the price at which we are indifferent to whether the customer forgoes the deposit for the convenience of wasting his bottle or returns it to get his money back.

One other way of deciding on the deposit is to ask at what price people would be indifferent about whether to return the bottle or not. A possibility is that discarded bottles, at a certain price, will be profitably collected by people who retrieve them and bring them to the collection stations. If at 15 cents all the bottles show up, we can excuse the people who leave bottles around, because by forfeiting their deposits they have paid for the cleanup. And if the bottles are not all retrieved but we can afford cleanup crews with the unclaimed deposits, again we may not care whether people return their bottles or pay in advance for the cleanup. These possibilities suggest the kind of deposit we might characterize as a price and why, having set a price, we might be content to let everybody decide for himself whether to pay the price and waste the bottle or save the price and return it.

This discussion does not settle how we should deal with the bottle problem; it only indicates what distinguishes a price system from other modes of regulation, including others that entail payments of money. Specifically, there is no implication here that we have to be ethically neutral about social behavior with beer and soda bottles. Also, there is no *a priori* assurance that any moderate price would cover the cost of cleanup. Furthermore, we may be attracted to a high deposit if the proceeds cover more than cleanup costs and leave us a balance for other purposes, especially if the people who forfeit deposits are people we do not mind taxing. If cleanup proves impossible, it may be hard to agree on the size of the indemnity to the local community that we would like people to pay in the form of forfeited deposits. And in all of this our evaluation may depend on how the forfeited deposits may legally be spent.

Allocating a fixed supply   Taxi medallions have counterparts wherever overcrowding is a problem. Taxis may be limited for the benefit of taxi owners, with results not altogether in the interests of their passengers, but the same principle of "marketable rights" has application to congestion or any threatened overuse of some limited capacity or facility. Parking is an example.

With medallions, there is a decision by a central authority on how many cabs might operate, but the question of whose cabs is left to the market. With downtown parking, a central authority may determine how many permits will be issued or how many spaces made available, or exactly where parking will be permitted, and then rather than choose who gets to park or where each person parks, or schedule individuals over the hours of the day or the days of the week, the authorities can let prices take care of those decisions. There are numerous ways. A familiar one is to install meters with rates low enough to keep spaces from going unused but high enough not to attract drivers in numbers much greater than the spaces available. If a uniform meter price leaves outlying spaces unused and centrally located spaces inadequate, higher prices can be set on the meters where spaces are in greater demand.

Alternatively, where people regularly park all day every day, windshield stickers can be sold—again at a price not so high that spaces go unused or so low that too many stickers are demanded, leaving favoritism by issuing clerks to determine which drivers get the stickers. (If the stickers are permanent, it is important that a person who no longer needs one, or would rather have the money than the parking privilege, be able to turn his sticker in for cash or to sell it.)

The price of parking does not in this case cover the cost of providing parking, but rather allocates a fixed supply. It ensures that spaces go to those to whom they are worth the most in money. The appeal of the system may therefore depend on what is done with the proceeds. The advantages of the system are the following:

It makes parking more orderly and eliminates congestion.

The costs of searching and waiting—costs that are "paid" by drivers but not received by any authorities—are minimized.

People can make plans and arrive on time.

The system is impersonal; there are no favors and no bureaucracy.

Those who get the spaces pay for them, and those who do not (to whom the spaces were evidently worth less than the money others were willing to pay) can in principle be beneficiaries of whatever is done with the proceeds.

Even the resident of a side street in a town that assigns each resident a parking space on his own street might wish in principle to have a meter installed in his own name so that the space need not go unused when his own car is elsewhere, and so that, if he would rather make other arrangements (such as disposing of his car and using taxis instead) and enjoy the profits from his meter, the space would be more productive.

Thus, the price attached to a parking space can be taken as an approximate measure of what some additional spaces would be worth. If the current meter or sticker rate is enough to cover the cost of additional parking, that is a signal that additional parking can be provided at no net cost to the authorities. If parking areas can be sold or leased for more than the net return from meters and stickers, somebody else has a more valuable use for the parking space, and the town is losing money by pricing the parking at less than the land is worth. If permanent parking medallions were issued, and could not be recalled without breach of faith, the arithmetic indicates that the town could cover the cost of buying back enough medallions to retrieve the land from parking and make it available for alternative use.

Discrimination by price   Parking spaces go to the people who can afford them. Whether their parking needs are especially urgent, or driving and parking saves them more money than it saves others, or they are merely wealthier, they get the spaces because they are willing to pay. The apparent discrimination is moderated by the fact that the wealthier the parkers are, the higher the price may go and the more the town can collect. We may nevertheless be tempted to offer spaces free, or at reduced rates, to the poor. A relevant if not decisive consideration should be that because they are poor they would evidently rather have the money than the parking. We might give them parking tokens and expect them to trade them in for cash. If we give them a "property right" that they can enjoy only by driving and parking, they have something less valuable than if they could trade it at a price at which they would rather have the money. There may be good reasons why that could not be done, and why neither the equivalent cash nor the transferable right is feasible; nevertheless, providing parking to someone who would not or could not pay for it is providing something the beneficiary values less than a share in the proceeds from an indiscriminate pricing system.

In closing this extended discussion of bottles and parking, which has been intended only to illuminate what is meant by *prices*, I acknowledge that these examples were picked because they are ideal illustrations that epitomize a wide range of activities to which the same principles apply. However, they are not typical. They are easy cases. Automobiles are large, easily recognizable objects; each is a discrete unit; each has an owner of record; cars are never confused with bicycles; they arrive and depart under their own power; tamperproof meters can be bought for a price that makes them a good investment; spaces are locally interchangeable and nobody needs to care which space within a block he occupies; and the money involved is not worth arguing about. Applying the same principle to pets in a public park might not be feasible. Similarly with the bottles. In the abstract, the principle applies to Kleenex, cigarette butts, candy wrappers, and eggshells, but we don't expect it to work for them.

Pricing Systems

For an economist, markets and prices play a central role. Environmental effects are effects that escape the market. They are outside the price system and "external" to the accounts of the participants. The economist diagnoses environmental problems, therefore, as a failure of the pricing system or the market.

People concerned with information, education, morals, civil law or law enforcement, and norms, traditions, and informal social controls characterize or diagnose environmental problems from their own perspectives. They see the problems as failures of information, of education, of self-control and public-spiritedness, of administration, of social norms, or of the legal system. Each emphasizes the absence of whatever his own profession looks to in the exercise of social control. More than that, each will seek a solution by repairing or extending or creating a price system, an information system, an administrative system, an adjudicatory system, or a system of ethics and morality according to what his own discipline understands best. There is nothing necessarily wrong with that unless each supposes his own perspective to be uniquely correct. They cannot all be right.

Two things can be said confidently about extending the price system as a way of bringing environmental problems under control. One is that there is indubitably a role for prices. Numerous successes can be pointed out. A good part of the development of economic institutions has been bringing what used to be externalities into the market system by the conferring and protection of property rights and the enforcement of contracts, by the availability of better information, and by technological development that facilitates marketing. The second is that some environmental problems are bound to remain beyond the reach of even the most ingenious efforts to create or extend markets. Indeed, there is a presumption that it is going to be difficult to bring additional effects into the market system; if it were easy, it would have already been done.

It is not enough to propose that airlines be charged for the noise aircraft make on takeoff and landing. There has to be a way to measure the noises made by different aircraft and to stipulate what the price should be, what to do with the proceeds, what authority will enforce the price system, what new safety or air-pollution problems may emerge as airlines reduce noise to reduce costs, how the local property values and taxes will change, and how fares and travel will be affected.

There is great variety in the pricing methods that can be and have been used in regulating activities that affect the environment. In recent years, attention has focused on what are usually called *effluent charges* or *emission charges* (largely because of the dominant importance of air and water pollution from stationary sources, partly because of specific proposals that became controversial). Pricing for environmental protection is not wholly or even mainly about emission charges, but they are currently the most familiar, controversial, and important policy alternatives.

Emission charges are furthermore a prototype for examination, despite the risk of reinforcing the misapprehension that they represent everything that pricing for environmental protection is about. Most of the questions about pricing (including the always crucial question "What alternative do you have in mind?") can be discussed in connection with emission charges. Most of the claims that can be made for the advantages of pricing can be illustrated, as well as the weaknesses and objections. So an examination of just what makes pricing mechanisms attractive to some people and unattractive to others, when and where they can be expected to work well or poorly or not at all, and where and when they have advantages over other more centralized and "regulatory" methods will emerge from an examination of the pricing of emissions.

I begin by exploring the different kinds of emission charges—what they are expected to accomplish, how they are imposed and administered, what activities and decisions are intended to be influenced by the charges, how the charges ought to be determined, and what alternative methods of regulation they should be compared with—and by straightening out the language. An *effluent* is something that flows out or flows forth, according to the dictionary, such as the outflow of a sewer, a drainpipe, a storage tank, or a pipeline. In environmental matters, effluents include smoke and gas as well as liquids. The term is readily extended to become synonymous with *emission*, which includes radiation, particulate matter, and even biological organisms and sometimes solid waste. In other words, almost anything that goes down the drain, up the stack, through the walls, or out the window, or that comes out the exhaust pipe, will be called an emission or an effluent. And the generic term *emission* then extends to aircraft noise, electromagnetic radiation, and heat from a cooling tower. Often (but not always—exhaust fumes and engine noise have been mentioned), stationary sources are implied. *Emission* or *effluent* also implies a regular product or byproduct of some process in which the emitted substance is an expected consequence; oil spills and explosions are not generally connoted. These terms exclude things that are carted away in containers or otherwise under control; slag, sludge, garbage, and spent reactor fuel conducted away by train, truck, or conveyor belt, and things safely contained underground or in storage structures, are not referred to as emissions. The image is of something that escapes containment and diffuses into the atmosphere or water or is otherwise broadcast, disseminated, or let loose. *Emission* and *effluent* are generic rather than specific. *Auto emissions*, for example, can refer to everything that escapes from tailpipe or engine or just to the particular exhaust components that are candidates for control. The *effluent* from an anchored fishing boat may be everything flushed into the water or only what is harmful or undesirable. *Smokestack emissions* may mean all the hot gases, or only those that do harm, or only those that have been identified as harmful or targeted for control.

Assume for illustration a charge that is not conceived as a source of revenue or as a penalty for violating some regulation. The charge is intended as a cost to the firm or individual (to simplify the image and the terminology, think of a plant belonging to a business firm) that attaches to one of the "outputs." Firms usually pay for their inputs: materials, fuel, electricity, labor, repair services, and the like. Sometimes they pay taxes on their output. They pay property taxes on fixed installations and they pay fees to operate. An effluent or emission charge is simply another cost that attaches to their production—this one a cost on the amount of some output the firm produces. It is an output, however, to which ordinarily no value attaches, or not enough to make the substance worth recapturing. The output or outflow is usually an unavoidable byproduct.

The alternatives to the charge are to leave the emission unregulated and to regulate it by what are usually called "regulatory standards." The alternative kinds of regulatory standards might be

• a limit on total emissions per day or per year,

• a limit on the emission per unit of legitimate input or output, such as on the sulfur emitted per kilowatt-hour produced, per million BTU of heat used, or per ton of coal burned,

• a prescribed reduction in total emission, or emission per unit, below some traditional level,

• a stipulated change in some input, such as pounds of sulfur per ton of coal or oil consumed, and

• a prescribed change in the mode of combustion or the equipment employed.

These might be varied according to season or wind direction or current air quality, if short-run variation is economical enough to bother with. And the limits may or may not vary with the plant's location; that is, plants located where more people or more crops or more wild animals are affected, or where the people and the crops and the wildlife are more vulnerable, may or may not be regulated differently.

The first thing to ask about a charge system is what the charge is imposed on. A second is what determines the amount of the charge (how many dollars per pound or per cubic foot of something). A third is how the charge will be administered. The object of the charge is to change the quantity, quality, or distribution by time and place of the emission. At the outset we can assume that the desired effects on emissions are basically the same with the charge system as with one of the other kinds of regulation. That is, if we had wanted, with alternative methods of control, to reduce total emissions, to even out emissions over time to prevent peaking of airborne concentrations, to induce relocation of plants downwind or downwater from population centers, or to stimulate new technologies that produced cleaner emissions, those are still central objectives with the charge system. The objectives need not be absolutely identical; there will be things that could be done with the proceeds of a charge system that, though they could be done with budgeted funds under an alternative regulatory regime, would receive different emphasis. But generally we should assume that there is some environmental problem to whose solution the charges and the other regulatory methods are alternative means.

There will be differences in strategy, because in the one case we are dealing with economic markets. An example might be the number of firms. Suppose we are dealing with an industrial location and an emission that has harmful effects over a local area, so that in principle the controls can be administered locally. What difference will it make whether there is one firm, or two, or ten, or a hundred? There could be a single large electric utility, two chemical plants, a dozen firms in the same light industry, or a hundred gasoline stations or laundries. There might be negotiation between the regulating agency and the regulated firms, and individual firms might keep an eye on other firms to get an idea of the rigor with which regulations are being enforced. But suppose a regulating agency were to limit emissions by auctioning permits, each permit allowing some quantity per month. Auctions work very differently according to whether there is a sole buyer, two or three who are both rivals and associates, ten organized in a trade association, or a hundred service stations. A single buyer at an auction places a low bid and takes everything; two or three buyers may collude on low bids, or compete fiercely to put each other out of business, or see an emissions quota as a way of sharing markets and restraining competition in a way the antitrust laws cannot prevent. A hundred service stations may compete and drive the price up to where, at that cost of emitting the substance, further output would be unprofitable.

So the way a charge system works will depend, among other things, on market structure. In general, characteristics of the activity that matter little for other modes of regulation may matter much for charge systems, and vice versa.

For simplicity, imagine a large number of plants (not necessarily identical in what they produce and how they produce it) that all emit a noxious substance. The emissions of each plant are susceptible to measurement, but once in the air the emissions of all the plants become mixed. Damage is not attributable to individual firms, but the emissions of every plant affect the total. How would a charge system work?

This is an easy case. We have an identified culprit (the noxious emission). We can monitor each plant. Only the total of the emissions matters, since the emissions are alike and commingled. All the plants are within a single jurisdiction, the same one in which the victim population resides. Our primary objective is to reduce total emissions, preferably in a manner that treats individual sources "fairly." More accurate, we want to limit the harm the emissions do; but we do that by reducing the emissions.

There are two alternatives to consider in what we know about the "damage function," about the way the severity of the damages relates to the total quantity of the noxious gas. One is that we know how much damage a unit of that noxious substance does, and we can put a money value on that damage. If we can value the damage caused by a pound of the substance without stipulating what the total is to which that pound is added, the damage function must be linear—proportionate to the total quantity. In that case we will probaby want to relate the charge to the damage, perhaps charging each plant for its contribution to the damage. The other possibility is that we begin with an idea of how much of the substance we are willing to allow and then set a charge that keeps total emissions within that limit. A possibility is that damages remain moderate up to a certain concentration and thereafter mount rapidly ("nonlinearly")—perhaps after a certain concentration is reached the atmosphere can no longer be refreshed by wind and rain. Or possibly, without any knowledge of how to assess damage, a target (an aggregate allowed quantity) has simply been set, and we want whatever price will get emissions down within that predetermined limit. The price that goes with that quantity can then be estimated. Permits can be issued at that price, and if the price turns out to be too high or too low to generate that demand for permits, the price can be adjusted up or down.

Alternatively, an auction can dispose of the target number of permits. The auction is merely a way of finding the price that corresponds to that quantity of emissions. Once the price has been found, if conditions do not change abruptly the auction can be dispensed with. Prices can be adjusted up or down when the demand for emission permits exceeds or falls short of the target quantity.

In principle there should be no difference between setting a price and letting demand determine the quantity of emissions and setting a quantity and letting demand determine the price to be paid. If we knew the demand curve—the schedule of prices corresponding to different quantities, or that of quantities corresponding to different prices (which is the same thing)—we would simply pick a price-quantity combination. The difference between the two methods of price setting is significant only when the price-quantity relation is uncertain. At the outset, if one is surer of the correct quantity than of the associated price, the auction-type technique is helpful; if one is surer of the appropriate price than of the quantity that would be allowed, the price can be stipulated and the quantity allowed to emerge. When damages are nonlinear in the total of emissions and rise rapidly beyond some concentration, setting the right quantity (short of the rapid rise in damage) will be the main interest. When damages are proportionate to total emissions, a predetermined limit will usually be of less interest than finding the right level of charges.

Now let us see how the result with a charge system differs from that with the more directly regulatory regime.

In both cases emissions have been reduced. One conspicuous difference is that in one case funds have been collected that are equivalent to the permit price times the remaining quantity of emissions. (And if the price corresponds to an estimate of the damage per pound of emission, the funds are equivalent to the estimated damage.) We will consider in a moment what might usefully be done with those funds. There is also the question of what may legally be done with them— whether they revert to the treasury, are directed by law to particular uses, or go into some environmental-protection budget.

Are emissions in the aggregate reduced the same amount? The answer depends quite simply on whether, if a quantity was determined and the price was allowed to emerge in the market, the quantity was the same that would have been allowed under a regulatory regime. If the basic determination was not of a quantity but of a price related to the estimated damage per pound, and if under the regulatory regime the regulatory standards would not have been arrived at through an implicit calculation of the quantity up to which the economic benefits to producers and consumers would equal or exceed the estimated damages, the quantity may well be different. It may be larger or it may be smaller. And if the regulatory regime would have established not a ceiling on the total amount of emissions but some ratio of allowed emissions per unit of product or per unit of activity or per unit combustion of fuel, we would have to examine and predict the reactions of individual firms under those regulations to arrive at an estimate of what the quantity might have been under the alternative regulatory regime.

Several things will be different. A striking difference will likely be that emissions produced by individual firms will not be the same. Indeed, a key result of the charge system is that it concentrates reductions among the firms best able to reduce emissions, since unit costs are higher for the firms least able to get emissions down.

Another likely difference is that the techniques of abatement or the methods of production that give rise to the emission in the first place may differ. Under the charge system each firm is free to reduce emissions the cheapest way it knows how. Under a regulatory regime the technology for abatement might have been specified.

A related difference in the overall result is that under the charge system less of the result was determined by regulation, more by decision of individual firms.

Will firms like the charge system better than the less flexible regulatory procedure? The answer is in several parts. Some firms may like it better and others not. Much depends on the markets in which firms sell their products, and on whether the firms are in competition with each other. Whether firms prefer charges will depend on the shape of the abatement function—not only how costly it is to reduce emissions by a small amount, but how much the cost per unit abatement rises as emissions are further reduced. Another part of the answer is that it will depend on what is done with the money collected. Finally, it will depend on whether the total of allowed emissions is more or less than would have been allowed under the alternative regime. It is instructive to look at these different parts of the answer.

That some firms will prefer a charge system and some will not is easily demonstrated. Suppose emissions with the charge are half the original quantity and the alternative would have been that every firm would have had to reduce emission by half. Firms will reduce emissions if that is cheaper than paying the price and will pay the price if that is cheaper than reducing emissions. Specifically, each firm can be presumed to have reduced emissions as long as reduction was cheaper than paying the price, and to have settled at a level of abatement at which further reduction would have been more costly than the permission to abate. Since the price charged for emissions is (we are assuming here) uniform among firms, all those emissions will have ceased whose elimination was cheaper than the price. So the aggregate abatement will have been achieved at the lowest possible total cost. Firms for which abatement was less costly will have reduced emissions rather than pay; firms for which abatement was more costly will have paid. Costs will have risen least in the firms able to abate most cheaply—precisely the firms that under a uniform 50 percent reduction would have done it at least cost. Any firm having reduced emissions more than 50 percent will have incurred the costs of reducing by 50 percent and in addition the costs of some further abatement, and will be paying charges only on the remaining emissions. So the cost to that firm will be more than the cost of 50 percent reduction but less than the cost to other firms that found abatement so costly they are paying and emitting at more than half the original rate. Firms that could have abated 50 percent only at much higher costs may be better off than at 50 percent abatement, even though they pay the charge on 100 percent of their emissions.

As mentioned above, the preferences of firms for the pricing technique might depend on what was done with the proceeds. It is easily seen that if the proceeds could be returned in some fashion to the firms or used for their benefit, they will collectively have saved money compared with the no-charge scheme in which each firm had to abate 50 percent. This follows from the fact that abatement costs in total were less than if every firm had had to effect a 50 percent reduction. On average the reduction was 50 percent, but most of the reduction occurred in the firms where reduction was cheaper than average. The total cost, the sum of abatement costs plus the price on remaining emissions, can be either more or less than it would have been under a uniform 50 percent reduction. (If there are many firms for which such a reduction would have been prohibitive, paying the charge may be so much cheaper for those firms that the total for all firms is substantially less than in the other case. On the other hand, if the abatement functions for all firms are sufficiently similar that they all reduced emissions approximately 50 percent, they have all paid about the same abatement costs as under the other regime, and in addition they are paying the charge on the remaining 50 percent.) Thus, any excess cost of the pricing scheme over the uniform 50 percent reduction is less than the funds collected (if the costs of administering the system did not eat deeply into the funds).

The money could not simply be refunded, or firms anticipating the refund would treat the charges as zero. But enough money would have been collected in emission charges to make it possible to reduce some non-emission-related tax to which the firms were subject. This would not interfere with the price incentive, and on balance the firms collectively would come out ahead compared with the uniform 50 percent alternative. There is no magic in this; it follows simply from the fact that the pricing system induces abatement by those firms that can accomplish it most cheaply. The firms collectively reduce the cost of abatement, saving money. So it is no surprise that any excess total cost, due to actual payments, is exceeded by the money collected.

The final point was that firms may prefer the pricing scheme or not according as the total reduction in emissions is more or less than it otherwise would have been. There are two important reasons—pointing in opposite directions—why the pricing scheme would lead to a different total abatement. Both reasons are based on the idea that any determination of how much emissions ought to be reduced is bound to depend on some estimate of how costly it will be to reduce them. Whether or not there is an explicit comparison of costs and benefits, any reduced level of emissions is a target for further reduction if cost is no object. Consider the simple "linear" case: If there were some substance that was harmful in proportion to its total quantity, with no "threshold" below which damages were negligible, so that the same amount of damage was averted for every pound eliminated from the atmosphere, and if it could indeed be eliminated at the source at a constant cost (with the final 10 percent of reduction costing no more than the first 10 percent), it is hard to see why any abatement program would stop short of 100 percent. If to begin with it is worth x dollars to eliminate y in damage, and if you can eliminate another y in damage for another x dollars, it is still the same bargain. If instead the first 10 or 20 percent of damage is easily abated but the next 10 or 20 percent is expensive, the third 10 or 20 percent nearly exorbitant, and the final 40 percent astronomical in cost, and if the commodity produced is one that cannot easily be dispensed with, so that consumers would be paying an enormous "pollution tax" in the elevated price of the commodity whose production costs were so increased, there would surely be a balancing somewhere, even if the environmental damages were judged quite severe. Thus implicitly or explicitly the costs of abatement are almost sure to matter.

And the pricing scheme yields information about those costs. Indeed it produces that information free of charge, without the need for regulatory agencies to do engineering or accounting studies. At whatever level of abatement the pricing system leads to, the price will measure the marginal cost of further abatement as seen by the firms. Up to that point abatement has been cheaper than the price, and beyond it would be more expensive. The price is an upper limit to the cost of the abatement already achieved and a lower limit to the cost of any further abatement. Whatever cost the regulatory authorities had in mind (or had calculated), the pricing system will display for them the marginal cost—the going cost for an increment of further abatement—as actually perceived by the firms. This is the nearest thing to the "real" costs that the authorities are likely to get.

Now there are three possibilities. One is that the 50 percent aggregate abatement is achieved at about the cost the authorities had in mind; they consider the abatement already achieved well worth the cost they are imposing on consumers of the products of these firms but further abatement too expensive to impose. A second possibility is that the cost is far higher than they anticipated, 50 percent abatement proving not nearly as achievable economically as they had hoped. The firms' customers will be paying a far higher "abatement tax" in the prices they pay than the authorities intended to impose on them. In this case the authorities may change their abatement target, lowering the price to one at which they think the smaller reduction in emissions is worth the cost. The third possibility is that the 50 percent abatement is achieved at a price far lower than the authorities ever dreamed of. If they had been thinking it would cost $10 a pound to reduce emissions of a substance by 50 percent, and after 50 percent reduction firms are paying only $1 a pound for permits on the other 50 percent, evidently they aimed far short of the abatement that can be achieved at reasonable cost. They should either reduce the supply of permits offered and let the price go up or (what is the same thing) raise the charge per unit emission and see the emissions go down.

The significance of the last point goes well beyond the original question — whether firms should be expected to prefer the price mechanism or the other kinds of regulation. Evidently, if the firms think the regulatory authorities underestimate the costs of abatement and may be about to impose an exorbitantly costly reduction on the firms, the pricing mechanism may get their point across credibly. They could simply challenge the authorities to stipulate the price that the authorities thought was a reasonable estimate at 50 percent reduction, and the firms would then continue their high emissions level only if indeed the abatement costs were higher than the price the authorities had stipulated. If on the other hand the authorities are too pessimistic, or have been cajoled by the firms into believing that abatement is terribly costly, the pricing mechanism is a good way of calling the firms' bluff in the marketplace. The firms might prefer not to reveal in the marketplace their true cost of abatement, living instead with a mandatory 50 percent uniform reduction that they could pretend was about as onerous as they could tolerate.

To conclude this latter point: It would be a mistake to think that the degree of abatement ought in principle to be independent of the choice of regulatory technique. Not only can any given degree of abatement be achieved at a lower aggregate cost by a scheme (such as a pricing scheme) that induces those firms to reduce emissions most that can do it most cheaply; it is equally important to note that the pricing scheme makes manifest the costs of further abatement (and the savings from a relaxation) at the margin. Whenever the direct estimates of abatement costs are uncertain, as they almost always will be, the information produced by a pricing scheme will be directly relevant to setting the environmental target.

The Rationale for Charging

There are several arguments for putting a price on harmful emissions, and, although they are compatible and reinforce each other, they are quite separate.

The market test

As noted above, a price offers an automatic adjustment process—a decentralized decision process—not only inducing the abatement in those particular firms, uses, or production processes where abatement can most economically be carried the furthest, but also providing a measure and a "market test." All abatement that is cheaper than the price will tend to be undertaken. Abatement that is so expensive that paying the price is preferable will be avoided, but only at a cost (paying the price) no less than what any other firm is expending on actual abatement. If the price has been determined at a level that represents the economic worth of abatement (that is, if it measures the economic damages that will be produced or the costs externally inflicted by the emission in question), each firm will undertake whatever abatement saves more damage than it costs, and no abatement will be undertaken that costs more than the damage is worth.

This argument is bound to carry weight if the damage can in principle be estimated and a value assessed for it—a value that can be made commensurate with the economic costs of abatement by attaching a money value per unit of emission. That the damage figure is not known (though it might be known in principle), and the estimates are very uncertain, docs not detract from the argument. As long as some estimate or conjecture of the nature and severity of the damage has to underlie any regulatory regime, it makes sense to induce the abatement that can be accomplished at a cost less than the damage, and to avoid abatement that would cost more than avoiding the damage is worth.

There are at least two fundamental problems with the above argument. First, if it is denied in principle that the harm and harmfulness of the emission can be gathered into a damage estimate and expressed as a price per unit of emissions, the argument loses not only its appeal but its basis. This is not the counterargument that what is harmed is "priceless" and that abatement should be unstinting and without regard to cost comparisons; if the damage is infinitely large, a prohibitive price is put on the emissions, and emissions cease—if necessary by cessation of the entire process that gave rise to the emissions. Rather, this counterargument is that the harm goes beyond the damage. It includes attitudes, responsibilities, and the government's philosophy toward the environment. It can include an interest in "criminalizing" certain environmental insults. It may reflect a belief that the real or imagined victims of environmentally transmitted damage, or of harm to the environment itself, will resent and distrust a policy that appears to make emissions optional, to leave them for the marketplace, or to merchandise abatement by getting it done by whoever can do it most cheaply.

A second and altogether different objection to the "market test" is that the costs and damages, even if perfectly measurable in dollars, cannot properly be added together. They cannot be minimized together because they accrue to different people. If my factory is causing a farmer (a wholly unrelated individual, possibly unknown to me) increased costs, reduced productivity, or damages that have to be repaired, at the rate of $1,000 a month, and I can reduce that damage by half at a cost of less than $500 to myself, but reducing damage another 10 percent would cost me substantially more than $100, why should I be obliged to clean up the first 50 percent and not further obligated with respect to the other half of the farmer's damage? True, together we are richer if I abate the first half and stop there. But there is, in the language of economics, a distributional dimension of the problem. This makes it inappropriate to do the arithmetic as though my factory and his farm were both parts of the same enterprise and together the two parts made more money if abatement went just up to the line where costs and benefits were balanced.

At this point the proponent of that market-test argument can retreat a little. Since the virtue of setting a price equivalent to the damage has not been appreciated on distributional grounds, he can still argue that the pricing approach, using a price that is not merely the estimated damage but a higher price if you please, will minimize the cost of whatever level of abatement is decided on by allowing it to be done by all those firms that can abate more economically than they can pay the price. Whatever level of abatement might have been achievable through some other regulatory mode could still be achieved with a price mechanism, but if our sympathy for the victims of the pollution in question is greater than our sympathy for the firms' customers we can induce greater abatement by imposing a higher price. (And of course if we have little sympathy for the victims and much for the people whose cost of living will rise as the costs of abatement show up in the prices they pay, we can tilt the process the other way and set a price lower than the estimated damage.)

Paying the costs
The second and third arguments for putting a price on emissions relate not to the abatement accomplished, but to the continued charges that

are paid for the abatement not accomplished. One of the arguments is simply that firms ought to be required to pay the full costs of what they produce, including the costs they impose on the public. The offending emissions are the "externalities" that in the absence of regulation were not being paid for. If what goes up a firm's smokestack deteriorates the paint on my house, reduces the productivity of my farm, requires me to install air conditioning, or increases my veterinary bill, I am contributing at my own expense to the value of the goods the firm is producing. If the firm rented my house it would pay me, but if it deteriorates my house it does not. If I provide part of my crop as raw material the firm will pay me, but if it reduces my crop by emitting something harmful the cost to me is the same but the firm does not pay. And it ought to pay. Not necessarily to me, but still it ought to pay. That is the argument. As stated, the appeal is ethical rather than economic. It need not have any punitive flavor. The firm may be producing a perfectly good commodity that its customers need, it may be emitting a substance whose harm can be estimated, and it may be willing to pay for what it subtracts from value elsewhere in adding value to whatever it produces. If it does, it need not apologize.

This argument fits almost hand in glove with the market-test argument. One argument is that by paying the price on its continued emissions the firm is complying with the rule that you ought to pay for what you get, where "what you get" includes the benefits of inflicting expenses on the public. The second argument is that paying the price "keeps the firm honest"; the price paid is the guarantee that the firm has not evaded any economically reasonable opportunities to abate the substance. Thus the price (pursuant to one argument) induces abatement at the firm's expense up to where further abatement would cost more than the averted damage is worth, and (pursuant to the other argument) the firm meets its responsibilities by paying for the damage caused by its continuing emissions.

Lest the argument sound too neat, notice that the market-test argument concerns an allocative function that the price performs—getting abatement done where it can be done most economically, and up to where the costs are no longer less than the environmental costs averted—whereas the "pay for what you get" argument is concerned with equity. To illustrate with an extreme case: An activity causes damage, but is so costly to curtail and so essential that no abatement at all is expected. The price is then a pure transfer; it has no effect on behavior. Whether the people engaged in the activity should pay is more a casuistic than an economic question. We may even want to find out who they are before we insist they pay just for the principle of the thing.

Disposition of proceeds

Meanwhile, back at the agency, the money is coming in. Whose money is it? What should be done with it? Is it a pure byproduct of a regulatory scheme, embarrassing to an agency that has no budgetary authority to spend it and perhaps appearing to be a revenue bonus available to meet public needs?

Here is where a third argument for pricing enters. It is not the argument that pricing produces revenue; the goods produced by the emitting firms could just as well have been subjected to an excise tax if we had wanted the purchasers of those commodities to provide revenue to the government. And if we think of the money as punitive damages, we can be pleased to have transferred money from offenders to the government, like the fines collected from traffic violations; but that is not the attitude that goes with pricing in the first place. The argument instead is that these funds, generated from a regulatory mechanism that induces an "economical" level of emissions to continue, are equivalent to the costs or damages estimated to be inflicted by those emissions. They ought to be used to complete the transaction by which the emitting firms pay for what they use and what they do. Paying into a public treasury on account of emissions and paying *for* the emissions by paying to repair, indemnify, or forestall the damage are not the same. And getting the benefits to the victim is every bit as important as getting the money from the firm that produces the damage. That is the argument.

It will be the rare environmental externality that permits us, at no exorbitant administrative expense, to identify all the victims and assess their damages individually. We probably have a crude aggregate estimate of damage, and at the level of an individual victim there may be no reliable way of determining how much of any apparent costs is actually due to the regulated emissions. Still, in principle the distributional objection mentioned above could be mitigated if funds were dedicated to completing that transaction.

How would that be done? The answer depends on how we arrived at our estimate of damage in the first place.

Completing the transaction

If the damage assessment was the estimated cost of cleanup, removal, repair, and protective measures, so that it is the cost of defending against the emissions rather than allowing them to do their harm, there may be a strong presumption that the funds should be used for exactly that. Specifically, using our earlier example, if we put a 15¢ price on the "emission" of unreturned bottles by requiring a deposit on all bottles, and if the 15¢ represents the estimated cost per bottle of cleaning up roadsides and parks, exactly the same argument that made 15¢ the relevant figure is an argument for using the money to clean up the bottles whose estimated cleanup cost underlies the whole scheme. To the extent that cleanup, repair, protection are best handled by public authorities, the proceeds of the emissions charge can be used to finance the appropriate programs.

Alternatively, the damage estimate may have reflected the costs imposed on individuals—costs that cannot be averted or repaired economically by public programs. In principle, if the estimate was a good one and if the costs of administration are not too large, funds exist to indemnify the victims for the full costs that were inflicted on them. Whether they protect themselves with air conditioning, reimburse themselves for the losses to their crops, cover the cost of more frequent painting, or take the money as pure financial compensation for some discomfort or privation, they get full restitution if our damage estimate was a good one and lends itself to decomposition into its individual components.

The principle can be made sharp and clear. Whether it can be done in practice is another story. Whether the adjudicatory or assessment process can be carried out fairly and without exorbitant cost and whether the procedure would invite spurious claims and self-inflicted damage (as when people are overinsured against fire or burglary) are bound to be disturbing practical questions about the possibility of completing that transaction. But if (as with all public expenditure or tax programs) the "target efficiency" leaves something to be desired and we do not demand that every individual break even but only that a reasonable attempt be made to provide protection, repair, or compensatory benefits, it may be possible to meet the proposal in spirit if not to the letter.

To recapitulate: There are at least four distinct attractions, separable but compatible, to a charge system. One is to get the abatement allocated among firms so that if some have lower abatement costs than others the cost of any given abatement can be minimized. (This statement can be turned around to read " . . . so that for any given aggregate cost the greatest aggregate abatement can be achieved.") Second, if the price is an approximation of the estimated damage per unit of emission, it offers a market test, revealing the marginal cost of further abatement and inducing the undertaking of abatement where it is no longer less expensive than the costs inflicted by emissions. Third, whoever does the emitting, or whoever purchases the goods and services produced by the emitter, "beneficiary," or "offender," pays the full "social cost" of the goods and services so produced. And fourth, revenue is available that could in principle be used for repair, reparations, and protection, in an amount sufficient to cover the costs inflicted by the emissions if the price has been approximated to those costs (with the "costs" or "damages" augmented by the costs of administering the program).

Focus on damage
There is a further advantage to pricing that may be enjoyed even if pricing proves to be impracticable or insufficiently superior to be worth substituting if the more direct regulation is in place: It inescapably entails deciding on a price. It focuses attention on the quantitative characteristic of the problem and the proposed solution.

Determining the price is giving an answer to the question "how much?," and exactly what the price attaches to is an answer to "a price on what?"

It is easy to talk in a vague way about putting a price on what goes up an electric utility's chimney, but ultimately the idea has to be reduced to something concrete, like $5 per ton of carbon dioxide, sulfur, or lead, measured at the smokestack or 500 miles downwind, every day or on days when the wind blows from the north, or when the concentration downwind reaches a certain level. And if the price is to be an approximation of the costs and damages inflicted, we not only have to identify the damaging substance but we need to know how to measure the damage—first in the units or quantities in which the damage actually occurs or in the units and quantities of protective or cleanup measure and then in money values commensurate with the price. Inescapably, therefore, a pricing system focuses on the damage, on the economic measure of the damage (or the costs of averting or repairing the damage), and on the actual agent that produces the damage.

To people who are attracted to pricing in the first place, this inescapable focus on the nature and measure of the damage will appear to be merely a virtue implicit in what has already been said. But it has a nontrivial implication: It draws attention to the fact that damages are not proportionate to emissions. Damages associated with emissions may vary substantially from source to source. A price that is uniform per unit of damage will not be uniform per unit of emission. Sources differently located should pay different emission prices for the same substance emitted if a pound of the substance emitted at one location does more damage than a pound emitted at another location.

Why might the emissions differ in the damage they do? One possibility is that low concentrations are harmless, and up to some threshold the damage is negligible. An opposite reason is that beyond some level all the damage is done—the fish are dead, the oysters infectious, or the water undrinkable—and the marginal damage of further emissons is zero. A third possibility is that the emission depends on the weather for the damage it does, being aggravated photochemically or harmlessly rained out of the sky, so that its potency depends on regional climate, local weather, and season of the year. Finally, and most important, the populations and other resources at risk differ from location to location, in number, in value, and in vulnerability.

This brings us to the divisive question of whether we want to discriminate among target populations. If a harmful activity is going to take place somewhere, exposing a certain population to the risk of sickness or death, contributing to the deterioration of their homes or crops or businesses or recreational activities, or entailing defensive costs to individuals or to the firms they work in or the communities they live in, do we want to measure, for the purpose of controlling emissions, the aggregate amount of risk and damage at each location? Or do we want to ignore the fact that aggregate exposure and vulnerability are less for emissions at one location than for emissions at another?

Notice what hinges on this question. If we estimate the costs and damages specific to a particular location and price the offending substance according to the likely damage per unit of emission at each location, we will induce less abatement where exposure and vulnerability are least and more abatement where exposure and vulnerability are large. More than that, because the processes that produce the emission become less costly at one location than at another, any kind of production that is not permanently attached to a particular location (on account of proximity to materials or markets) will have some inducement to relocate where exposure and vulnerability are least. New plants, especially, may be located where the emission price is low, just as plants locate where property taxes, wage rates, or energy costs are low. And the higher the level of emissions per dollar of output, and the higher the cost of abating emissions, the greater will be the inducement to relocate where the emission price is low.

We are back to that argument that is appealing in the aggregate but poignantly divisive for individuals: There may be fewer people exposed in a particular location because the population is of low density, but each individual feels just as exposed as if he had more neighbors. If two populations are of equal size but one has more people who are especially vulnerable (for example, the elderly), the vulnerable members of the population that is more vulnerable in the aggregate "protect each other" by collectively making it an expensive place to emit the offending substance. And the few elderly or otherwise vulnerable people in the area where the damage that may be done to them is priced low (because they are few) may not appreciate feeling sacrificed for some collective greater good or lesser damage.

Whatever conclusion one reaches about the manifest benefits of discriminatory pricing or the patent unfairness of it, it is an issue that ought to be faced. Pricing schemes require facing it.

Schemes that focus only on the amount of sulfur emitted (irrespective of the target), or on the ambient concentration of a sulfur oxide (irrespective of the exposed populations), deal with a damaging substance but not with the damage. They control the agent but not the effect. Not being concerned with damage, they do not require its identification and measurement.

Pricing schemes, in contrast, require the identification and assessment of damage. They require facing up to questions exactly like this one: Do we want the abatement of noxious substances to be greatest where the exposed and vulnerable population is greatest and least where the exposed and vulnerable population is small, and do we want the emissions to be relocated so that the damage per exposed individual will vary from place to place but the damage for the whole population will be minimized?

For many of us I expect that the answer will be "It depends." If we are dealing with aggregate damage to livestock or forest products, or cleanup costs inflicted on homes and shops, or noise insulation made obligatory in schools and other public buildings, confining the potential damage to small towns and low population densities will seem eminently sensible. However, if we are dealing with carcinogenic or (especially) mutagenic substances, we may be more uneasy about merely maximizing damage abatement. We may even be uneasy about formulating the issue so that we have to face it explicitly.

Other Pricing Procedures

There are other pricing mechanisms. To show that charges on the offender are not the only possibility, we can propose the exact reverse: If we have no authority to command that the offending output be attenuated, or to impose a charge or penalty in support of reduction, offering a price may serve the purpose. If we have a damage estimate in dollars per pound of a substance emitted, and we cannot charge that price to the emitting firm, we can pay it that price per pound of emissions abated. One way we need police power, the other way we need purchasing power. The identical price ought to lead to the identical result, with the not-inconsiderable exception that the emitting firm is receiving rather than paying.

If $1 per pound has to be paid on all emissions, abatement will proceed to the point where it ceases to be cheaper per pound than the $1 that can alternatively be paid. But if reimbursement is available at $1 per pound of abatement, it is equally profitable to carry abatement to that same point where the cost per pound of abatement is no longer less than the price.

There are two differences. The obvious one is that the firm's financial position differs by a fixed amount, equal to the original emissions times the price. Compared with no price at all, it is ahead a fraction of that amount if it receives payment for abating, and it is behind by the complementary fraction if it pays on emissions instead. A related difference (possibly a second-order effect) is that in the less profitable situation production costs are higher; thus the retail price will go up and consumers will buy less. All this accords with what we expect when hidden costs are discovered. Costs go up, prices go up, the industry contracts, and the overall reduction in emissions (perhaps to be distinguished from the "abatement," where the latter is thought of as reduced emissions for the given activity rather than reduced activity) depends on the elasticity of the demand for the product and on how large the abatement costs are as a component of total cost.

The second difference is that to pay firms for abatement we need a baseline. There has to be some initial position from which reductions will be counted as abatement. When the firm pays on the basis of actual emissions, the baseline is usually zero and goes unnoticed. (If the emissions on which a price is based are only "excessive" emissions, we need some historical or technological standard; that possibility, which presents an exact counterpart to the present complication, was omitted above.)

Administratively this can be a problem. We may not have good base-period data on the firms. Or we may have data from the year or two before the scheme goes into effect; but if the firms anticipate being paid for reducing emissions they may boost emissions during that period. Even if that can be forestalled, there is the question whether each firm should be allowed its own individual base, especially since that gives the more offending firms more room to be rewarded for merely catching up.

A technological norm might be established, but that entails two difficulties. First, firms that emit the same substance are not necessarily producing the same commodities: The same substance may simultaneously be emitted in wood finishing, paint thinning, leather tanning, printing and engraving, and a variety of other activities, making it hard to establish a common normal emission ratio. Second, even though no opprobrium ought to attach to a high level of emissions that was not illegal or improper until an abatement scheme went into effect, the potential windfall gains of the firms that can abate most substantially are not likely to be regarded as altogether legitimate.

The problem generalizes. There are numerous instances in which people could be penalized or charged for a proscribed activity because it is comparatively easy to monitor the level of activity, where rewarding good behavior (if it is merely the absence of disapproved behavior) would be open-ended. Both a fine for littering and a charge on littering may be feasible, but how do you pay picnickers and passing cars for everything they did not leave behind or throw out the window? The conclusion is not that people and firms can never be paid for their abatement; there are instances where it actually works. The point is that there is an inherent asymmetry. There is a crucial baseline parameter to be determined, one that usually is conveniently missing from the more familiar case of charging the emitter.

These considerations are tangential to the question of who ought to pay or be paid—the question whether I have a right to play music or you have a right to quiet, whether I have a right to plow or you have a right to dust-free air downwind, whether I have a right to hunt migratory birds or you have a right to their sanctuary, whether I have a right to do wood finishing or you have a right to be free from the fumes and the fire hazard. Sometimes these Rights (with a capital R) are there to be discovered in the courts or in our consciences; sometimes rights (with a lower-case r) are to be decided by legislatures. There is nothing in principle that tells quite where to draw the line in questions such as whether I should be free to do something unsightly with my house in your neighborhood, whether I should be allowed to breathe freely in a public place if I have a respiratory infection, fishermen to disturb the quiet of a swimming place, infants to cry, or cats to be allowed out if they eat birds. There is much literature on the efficiency or costliness of assigning the rights to the plaintiff or to the defendant when civil suit is the mechanism of control. Here it is enough merely to notice that there is no universal presumption that the ethically correct way to induce a firm to reduce some harmful emission is to command or to entreat, to charge or to reimburse.

In paying for abatement rather than charging for emission we again have the two parts to the transaction and the question whether we wish to complete the transaction. To this point we examined the possibility that the emitter should be paid for emitting less. There is a corresponding question whether the beneficiary should pay for value received. A possibility is that nobody else will pay, and if the beneficiary wants the abatement, all he can do is pay. A price system may then let him pay up to the point where further abatement is not worth the price; and if the price reflects the emitter's costs of abatement, this would be the point at which the emitter and the beneficiary have minimized their joint costs. The earlier discussion considered the possibilities that the costs or the damages inflicted by the emissions would be centrally estimated as a basis for pricing, that the authorities might not know the costs of abatement, that a pricing system would generate information, and that the decentralized adaptation of firms would be an efficient response to the cost characteristics of individual firms. It is often (perhaps typically) the case that, however difficult it is to estimate abatement costs, it is even more difficult to attach a reliable value figure to the costs and damages averted by abatement. Pricing is one way of attempting to cope with that part of the problem too.

In particular cases of purely bilateral damage, in which I am the exclusive source of the emissions that hurt you and you are the only one hurt by my emissions, there may be a certain symmetry in the negotiated outcome no matter who has the initial rights—whether you have the right to be left unhurt (and I must quit or compensate) or I have the right to pursue the activity (and you must acquiesce or reimburse me). If my abatement costs exceed the damage you wish to avoid, in the one case I will pay damages rather than abate; in the other case you will take the damage rather than incur the greater expense of curtailing it. Only if it costs less than it is worth am I likely to avert rather than pay damages or are you likely to pay for relief. The negotiated transfer might be anywhere between those two limits, depending on whether, for example, you have the right to demand only damages or the right to forbid the activity unless I pay your price. (In that case I may pay any price up to the full cost of abatement.) We have here a kind of double market test. The costs on both sides of the transaction play a role in the decision.

The difficulty in the typical case is that my abatement benefits numerous parties downwind, downstream, or by the side of the road. Any beneficiary who refuses to pay his share can be a "free rider," and if nothing is done until everybody pays a share, anyone can insist that he doesn't mind the emissions and nobody should expect him to pay. If you let ragweed grow in your fields and all of your neighbors are prepared to pay the cost of suppressing ragweed, even somebody who suffers acutely from hay fever may pretend to be allergic to something else (even to the proposed pesticide) to be excused from paying his share and even to stake a claim for compensation. (And you may grow ragweed for the profit of being paid to suppress it!)

In the terminology of economics, abatement often has the characteristic of a "public good," something that if produced at all becomes available unstinted to all beneficiaries and cannot readily be withheld to exact payment. But if the water in a pond that had become unfit for swimming is cleaned up at great cost, it may be possible to deny access except to people who pay their share. That still has the problem of gratuitously excluding someone who likes to swim but not quite enough to pay the standard share; unable to document that he likes swimming less than I do, he forgoes it while I pay, and his potential enjoyment goes to waste.

Generally, then, charging beneficiaries is not as promising in practice as in principle it might sound. There is, though, an important case that was hinted at above and will show up in one of the case studies: If an entire area (perhaps a residential area) can be made free of some noxious element and the benefits can be observed as increases in market values (rents and sale prices), it may be possible, if the authority exists, to capture some of the windfalls. If rents double because aircraft noise is eliminated, somebody who does not mind the noise may claim to be unbenefited; but if property values go up he may take his gains and go to live where it is cheaper, converting the quiet into a capital gain. In this case the "public good" problem is still present but the basis for assessing gains may be less elusive.

Fleet mileage

An unusual and instructive example of a pricing mechanism, this one for both energy conservation and environmental protection, is the fleet-mileage standard established for passenger cars produced in the United States. It is unusual because it appears at first glance to be direct regulation of the most inflexible sort. Each year there is a numerical figure for miles per gallon that the manufacturer must meet or pay a penalty. There is no credit for being ahead of schedule or doing better than the prescribed figure; no carryover, permitting a manufacturer who excels in one year to average out the following year; no falling behind and making it up next year. The penalty is a fine; it is not a "price" that, once paid, entitles the manufacturer to a shortfall in performance.

This case is instructive because embedded in it is an interesting use of the price system. Indeed, the concept on which the fleet-mileage standards are built is a regulatory innovation. (Whether it has potential appliction other than in automobiles is not certain.) The principle is important. Automobiles are important. And in displaying the rationale for pricing, and showing how pricing can sometime be embedded in a more administrative scheme, the case is instructive.

The key is the averaging. The manufacturer is not given a mileage standard that every car must meet, or a schedule of mileages for cars of different weight, engine size, or price, or quotas by number or sales value of cars in alternative mileage categories. Rather, the manufacturer is given a mileage figure that all cars produced during the year must together meet on the average. The average is simply the arithmetic mean. By 1985 the average, under 1980 law, will be 27.5 miles per gallon. A manufacturer will be within the requirements if every car meets that figure. He will also be within the requirements if half the cars achieve 30 m.p.g. and the other half at least 25. If two-thirds achieve 30, the remaining third can be as low as 22.5. Cars can get as few as 10 or 15 m.p.g. as long as there are compensatory sales of cars that keep the average up.

Insofar as gasoline mileage is concerned, no individual car buyer is precluded from buying any kind of car. An automobile company can sell anyone a car that gets only 10 m.p.g. on the condition that it simultaneously sells seven cars that get 30, three that get 33.3, or one that gets 45. The snag is that for each of us who would like a car that gets only 10 m.p.g. there may not be somebody who will take a light-weight 45-m.p.g. car, or three who will buy cars getting 33.3, or seven who will buy cars that get 30.

This is where the price system comes in. If the cars that all the customers together would like to buy when cars are priced the way they are normally priced average out at 27.5 m.p.g. in 1985 (and the corresponding lower figures for the years before 1985), the regulation will be redundant. But if customers choose a mix of cars with an average mileage below the specified level, something will have to give. Dealers then will find that they can sell more low-mileage cars that they can get their hands on. They will offer fewer discounts and other customary inducements on those cars. Manufacturers will not, because of the fleet-average requirement, be able to keep up with the demand for the low-mileage cars. They will have to require dealers to take more high-mileage cars than they want, or make it attractive to dealers to sell more of the high-mileage cars than the customers want. In the end, what will keep the customers satisfied and meet the fleet-mileage standard will be price adjustments. To earn the high markups that are possible on low-mileage cars, the manufacturers will have to cut prices to sell the high-mileage cars that are not in sufficient demand.

The result is not much different from that of an excise tax on low-mileage cars to discourage their use (a "weight charge" or "energy utilization" charge) coupled with a negative tax (a subsidy) on energy-saving cars. Whether the purpose is energy conservation, environmental conservation, or both, one can design a combination of taxes on the cars to be discouraged and subsidies (leading to discounts) on the cars to be encouraged, using the taxes to finance the subsidies and breaking even in revenue. The buyers of the low-mileage cars thus pay an enticement to others to buy the high-mileage cars that "justify" the low-mileage cars.

The fleet-mileage standard, imposing only an average on the entire fleet of a given automobile producer, leads to pricing that will look like a combination of a premium (a tax) on cars that are in short supply because of the regulation and a discount (a subsidy) on cars that need to be sold to keep the average up. But it is up to each manufacturer to accomplish all this by adjusting prices on different models and makes and styles, meeting the competition from other automobile firms and from imports, and always selling cars to customers who are subject to no regulation themselves.

How effective the fleet-mileage standards have been is not easy to judge. Gasoline prices rose so much during the period in which the mileage standards were imposed that the market shifted in favor of high-mileage cars. In retrospect, gasoline prices may have done what the mileage standards were intended to do; the mileage standards may thus have been unnecessary. Furthermore, the gasoline prices worked in the same way: High-mileage cars became relatively cheaper over their lifetimes, if not at initial purchase, and low-mileage cars more expensive. (The mileage standards may have induced companies to plan on a higher-mileage fleet, and they may have been somewhat readier than they would have been for the market shifts that occurred.)

Converting nontransferable rights
In many areas of regulation, not only the environmental area, rights to participate in some activity or to use some resources evolve over time or are granted in accordance with traditional shares. Water rights and import quotas are examples. The imposition of air-quality standards over a local area, as under the Clean Air Act, has tended rather naturally to treat existing stationary sources in that fashion. Where existing sources together are within the required or implied level of total emissions they have been permitted to continue their individual emissions at traditional levels, not to increase them, while new firms or enlargements of existing firms must be accommodated within whatever room remains for further emissions within the established limits. Once the "excess capacity" is used up, there is no room for further growth and no room for new firms that might produce more and emit less than existing firms.

Proposals have arisen in all the three cases mentioned (California water rights, oil import permits, and emissions into the air from stationary sources) to commute those specific rights into ownership rights that can be traded, sold, or otherwise transferred or marketed. New or expanding firms could then buy their way into the market, obtaining the water rights or the emission rights from firms to which they had earlier accrued. Firms that could produce more value per unit of emissions would presumably value the emission rights more than some established firms, and a better utilization of the local air capacity could be obtained.

Objections are occasionally raised to the apparent "monetization" or granting of pure wealth where there earlier existed only a traditional right to participate. But what happens is that a valuable right, one whose value may have been unseen because it was never commuted to cash in the market, comes into view. Furthermore, it appears to become worth even more to the firms that elect to sell their traditional rights; otherwise they can always keep them. The new firms whose emissions replace the old emissions are advantaged by being allowed into business, air quality is unimpaired, and the apparent windfall gains have actually been created by enhancing the value of the capacity of the air to absorb emissions up to whatever level was already determined.

The principle is attractive. It already works for taxi medallions, could undoubtedly work for oil-import quotas, and appears feasible for a tangible scarce commodity like water or even some mineral rights. The particular characteristics of local air-quality regulation make practical application more problematic. One of the case studies is an examination of how complex and variegated these local "markets" in emission rights may be.

Case Study 1

The Regulation of Aircraft Noise
David Harrison, Jr.

Chapter 2

The Problem of Aircraft Noise

Aircraft noise is a major annoyance for six to seven million Americans. Although no one would argue against the benefits of jet travel, airport neighbors do question why they should bear the costs of this technological advance in the form of noise that disturbs sleep, interrupts conversations, and otherwise detracts from their environment. Government agencies and airport operators have responded to complaints in rather traditional ways, primarily by mandating noise standards for jet aircraft.

Economists have long criticized this "command-and-control" approach to regulation as wasteful and potentially ineffectual. They have advocated instead a strategy that would deal with major pollutants by instituting a market, either by setting prices that equal marginal social cost or by determining the socially optimal amount of pollution and selling rights to the highest bidders. Most of the literature on market approaches to pollution control has been in the context of air and water quality, but several authors have proposed such schemes for aircraft-noise abatement as well (Alexandre and Barde 1974; Council on Wage and Price Stability 1977; Muskin and Sorrentino 1977; Nelson 1978; Nierenberg 1978; Pearce 1976; Straszheim 1975; Suurland 1977).

Government regulators and noneconomists generally respond to noise charges or other economic-incentive proposals by admitting that the schemes may be desirable in theory, but dismissing them as unworkable in practice. Part of the reason for this criticism is that most analyses of economic-incentive schemes have addressed environmental problems at a somewhat abstract level, providing little information on how such schemes would be implemented. Economists have typically also given insufficient attention to the potential drawbacks of such strategies— particularly their information requirements—and to the importance of objectives other than efficiency.

The major objective of this evaluation of economic-incentive schemes to control aircraft noise is to provide a background for the task of translating theory about regulatory reform into better public policies.

Several recent studies have provided details on such schemes to control air and water pollutants (Hahn and Noll 1981; David et al. 1978; Mills and White 1978; Anderson et al. 1981; Palmer et al. 1981; and the two other case studies in this volume). The choice of aircraft noise for this case study relies on the fact that this form of pollution does not involve risks of death or serious illness, policy areas where experimentation is usually not possible. Noise may be a field where analysts can identify and explore the variety of considerations necessary to determine the best regulatory approach.

Another important concern analyzed here is the possibility of improving regulations by adding geographic diversity. The essence of the economist's argument for an emission charge is that it provides firms with the flexibility to reduce emissions in the least costly manner, both in the short run (when equipment and processes are fixed) and in the long run (when they have incentives to find new and cheaper ways of reducing emissions). Another means of achieving more efficient regulations is to allow environmental charges or standards to vary with the severity of the problem. Though recent papers have highlighted the theoretical advantages of geographic diversity, relatively few studies have empirically tested its importance in specific regulatory situations.[1] One of the objectives of the following analyses is to determine the advantages of noise regulations that vary by airport.

The remainder of this chapter provides a background of the technical, economic, and legal aspects of controlling aircraft noise. Although this material is lengthy, it is essential to any evaluation. Indeed, such a detailed analysis of "context" distinguishes this project from other studies. Some readers may prefer to skim this section and refer to it while studying the comparisons of regulatory approaches presented in the next two chapters.

Chapter 3 evaluates noise-control strategies at the federal level, focusing on comparisons between current noise standards for existing jet aircraft and noise-charge schemes. These comparisons relate to both the efficiency and the practicality of the two approaches. The chapter discusses why a national marketable-permit scheme is less attractive than a charge scheme for controlling noise, and includes an empirical analysis of the efficiency gain for adopting noise standards that vary geographically.

Noise charges or marketable permits could also be used by operators of local airports. Chapter 4 considers the advantages and disadvantages of a single airport switching to a noise charge. This discussion includes issues (such as the effect on airport traffic and regional employment) that may be particularly important to an airport operator and thus may determine the likelihood that he will establish a noise-charge scheme. Chapter 5 summarizes the major conclusions of the study.

Measurement of Aircraft Noise

Often defined as "unwanted sound," noise is measured in decibels (dB), a logarithmic transformation of sound energy. This relatively narrow scale represents the wide range in sound energy that the human ear can detect. Moderate speech heard at a distance of 3 feet is 60 dB. The sound on a subway platform when the train arrives is 90 dB. At 110 or 120 dB, the sound is intense enough to induce pain or at least a tickle in human ears.

The notion of "loudness" is, of course, subjective. Experiments to determine what decibel changes correspond to perceptions of sound increase have demonstrated that subjects will tend to increase a given sound by 10 dB when asked to double its loudness. Increasing a whisper heard at about 3 feet (40 dB) to a shout (80 dB), therefore, does not double the loudness, but rather increases it sixteenfold. These results mean that a given increase in decibels becomes more annoying as the decibel level is raised.

Aircraft noise is universally measured in terms of effective perceived noise decibels (EPNdB). The EPNdB scale includes several modifications to the basic decibel scale to account for how people judge the noisiness of an aircraft's landing or takeoff. A microphone would record an aircraft "flyby" as an increase in decibel level that reaches a peak and then gradually diminishes. One of the corrections in the EPNdB is to aggregate the noise over the entire flyby. In addition, the EPNdB scale incorporates the greater annoyance of high-frequency noises—which jets produce in relatively high proportion—by weighting these sounds more heavily. Because of evidence that a constant sound is particularly irritating, the scale also adds a correction factor from discrete (or pure) tones in the flyby's sound profile.

To compare the noisiness of different types of aircraft, the point at which the flyby is measured and the aircraft's gross weight must be specified. The noise-source controls established by the federal government measure an aircraft's EPNdB level at three points: landing—one nautical mile from the landing touchdown, directly under the aircraft path; takeoff—3.5 nautical miles from the brake release, directly under the aircraft path; and sideline—at the location of maximum noise along a line parallel to and at a distance of 0.35 nautical miles from the runway centerline (for aircraft with four or more engines) or 0.25 nautical miles (for aircraft with three or fewer engines).

Exposure to Aircraft Noise

The noise exposure forecast (NEF) measures the average noise value of aircraft operations over a 24-hour period, calculated for average conditions during the year and taking into account the noise from individual aircraft flyovers as well as their frequency. The NEF values include a correction factor that implies that one nighttime flight (occurring between 10 P.M. and 7 A.M.) is equivalent in annoyance to about twelve daytime flights.[2]

Noise exposure forecasts for airports usually range in five- or ten-unit increments from NEF 30 to NEF 45. The map presented in figure 2.1 displays the noise-exposure contours (places with the same NEF value) for Boston's Logan Airport. The distinctive pattern of the contours is, of course, due to the particular runway configuration and associated flight tracks.

To determine the number of persons subjected to different levels of aircraft noise, the NEF contours can be combined with population surveys. The estimates in table 2.1 indicate the number of persons residing within NEF 30, NEF 40, and NEF 45 noise contours of the 23 U.S. airports with the greatest noise problems. As these data demonstrate, aircraft-noise exposure is concentrated in a relatively small number of locations; most of the heavily affected areas are densely populated neighborhoods near airports built long before the development of jet aircraft and the expansion of air travel.

Valuation of Noise-Reduction Benefits

The NEF calculations are a well-developed and widely accepted basis for modeling the relationship between emissions and ambient-noise concentrations. To compare costs and benefits or to set a benefit-based noise charge, however, requires placing a monetary value on noise exposure. The value we seek is the willingness of noise-impacted households to pay to reduce their NEF levels. The task is complicated because the value placed on a given NEF decrease may vary with the household and the precontrol NEF level. Figure 2.2 shows alternatives for the

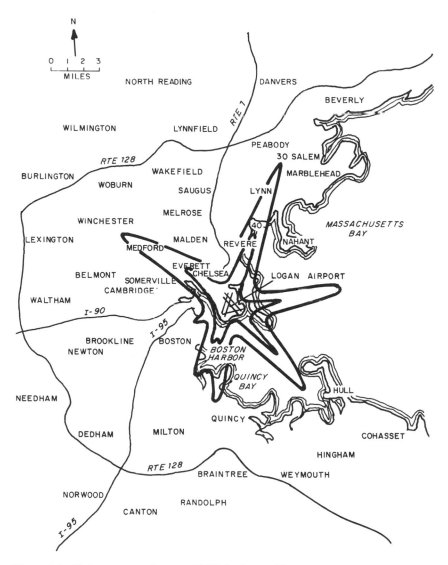

Figure 2.1   Noise exposure forecast (NEF) for Logan Airport, Boston.

Table 2.1 Populations inside various NEF contours around 23 airports in 1972.

	Thousands of people within contour		
	NEF 30	NEF 40	NEF 45
Atlanta	99.8	27.0	8.7
Buffalo	113.8	9.7	1.7
Cleveland	128.7	11.2	4.1
Denver	180.3	28.3	10.9
Dulles	3.5	0.0	0.0
J.F.K. (New York)	507.3	111.5	41.9
LaGuardia (New York)	1,057.0	17.1	0.0
Logan (Boston)	431.3	31.0	5.6
Los Angeles	293.4	51.1	28.8
Miami	260.0	29.7	6.4
Midway (Chicago)	38.5	1.8	1.0
Minneapolis–St. Paul	96.7	8.8	0.3
Newark	431.9	27.5	7.1
New Orleans	32.5	8.9	6.1
O'Hare (Chicago)	771.1	66.6	28.7
Philadelphia	76.9	0.3	0.0
Phoenix	20.5	6.2	1.4
Portland	1.2	0.3	0.0
San Diego	77.3	24.0	3.3
San Francisco	124.4	11.4	2.3
Seattle	123.2	17.3	3.4
St. Louis	100.0	8.5	1.7
Washington National	24.4	0.0	0.0

Source: U.S. Department of Transportation 1974, pp. 3-7–3-29.

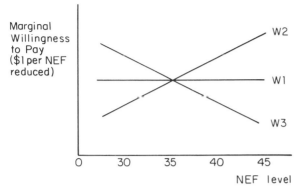

Figure 2.2   Possible willingness-to-pay functions.

function relating the willingness to pay for a small improvement in the NEF level to the precontrol level. In W1, the value a household places on a one-unit NEF reduction is the same whether the current level is very high (NEF 45) or reasonably low (NEF 30). In contrast, in W2 and W3 the value placed on noise reductions will vary with the household's current noise level.

The linear willingness-to-pay function provides the most convenient formulation for aggregating the value that households place on nonmarginal changes of the kind that would follow a major reduction in aircraft noise. One can aggregate noise-exposure reduction into the number of household-NEF units the change produces, and then multiply the total exposure reduction by the constant value that households place on a one-unit change. The benefit-based charge approach is also most appealing if the willingness to pay is constant. In that case, charging each airline a constant amount for each NEF-household its planes create will lead profit-maximizing airlines to produce the optimum level of noise reduction, and this optimal charge can be set without regard for either the current NEF level each household experiences or the costs that airlines incur in reducing aircraft noise. If marginal willingness to pay is increasing or decreasing sharply, it may be better to set standards.[3]

One technique for estimating the value people place on environmental improvements is to calculate the physical effects of the change (lower illness rates or more peace and quiet, for example), and then to determine the dollar value that people attribute to these effects. Alternatively, the willingness to pay for a better environment can be estimated from differentials in housing prices. This technique assumes that households are willing to pay more for an otherwise identical house in a less noisy location. The price differential set in the housing market therefore represents an estimate of households' willingness to pay for reductions in aircraft noise.

Effects of Aircraft Noise

Noise may damage hearing, interfere with speech, sleep, or other activities, and generate other detriments to human health. Loud and persistent noise, a risk commonly faced by industrial workers, can produce temporary and permanent hearing loss. In response, the Occupational Safety and Health Administration (OSHA) has established industrial noise standards that prohibit exposing workers to 90 dB for more than eight hours, or to 100 dB for more than two hours (29 CFR 1910.95).

There is some dispute whether airport noise—even at very high NEF levels—can cause significant hearing loss or other health-damaging effects. The Noise Control Act of 1972 requires the Environmental Protection Agency administrator to "publish information on the levels of environmental noise the attainment and maintenance of which in defined areas under various conditions are requisite to protect the public health and welfare with an adequate margin of safety." Using data on hearing loss among industrial workers, the EPA has estimated the level of community noise that might damage hearing assuming a 40-year exposure period (U.S. Environmental Protection Agency 1974). That study concludes that noise levels of roughly NEF 40 could result in hearing loss, measured as the possibility of incurring more than a 5-dB noise-induced permanent threshold shift.

As the EPA acknowledges, the assumptions in the study tend to produce a low boundary for hearing loss. For one thing, the calculations assume that equal sound energy causes equal harm, and thus that workers' hearing losses in short-term, high-noise situations can translate into losses from 40 years' exposure to much lower noise levels. The study also assumes that all aircraft noise is at the frequency at which the ear is most sensitive. In addition, the level protects the most sensitive 4 percent of the population from hearing loss; for 96 percent the noise level would have to be greater to induce any hearing damage. The measure of hearing loss was also set at a low level: a shift of 5 dB in hearing threshold, which is the minimal detectable difference in sound.

For all these reasons, one should be skeptical of the conclusion that those exposed to NEF 40 noise levels experience hearing loss.

There is no dispute, however, that aircraft noise is annoying. The results of an extensive examination of community responses to aircraft noise, summarized in figure 2.3, illustrate the relationship between NEF and the percentage of the population disturbed in such activities as sleep, relaxation, conversation, and television viewing. Federal Aviation Administration 1977 and other studies provide the basis for figure 2.4, which shows the percentage of population rating the noise associated with a given NEF level as acceptable or unacceptable.

Estimating Willingness to Pay from Housing-Price Differentials

When people buy a house or rent an apartment, they purchase the neighborhood's quality as well as the unit's structural characteristics, such as number of rooms or condition of plumbing. Neighborhood characteristics include accessibility to employment and shopping areas, proximity and quality of schools and other local services, property-tax rates, and other variables. For most airport neighbors, aircraft noise is the most important environmental factor. Since one can avoid noise by living outside the affected area, housing prices in noisy neighborhoods should be discounted, with the size of the discount reflecting the households' willingness to pay for lower noise.

There are a host of methodological and empirical problems in estimating the willingness to pay for environmental amenities from housing-price data (Harrison and Rubinfeld 1978a). The major conceptual issues involve the shape of the willingness-to-pay function, the possibility that willingness to pay depends upon a household's income and tastes, and the possibility that the market is not close to short-run equilibrium.

The conceptually correct procedure for determining the shape of the willingness-to-pay function would consist of two steps: (1) Estimate a general relationship between housing price and various structural and neighborhood characteristics; (2) use the resulting estimates of the willingness to pay for a small change in noise levels to determine the willingness-to-pay function. One could also test for the influence of income and taste variables on households' willingness to pay for noise reductions (Harrison and Rubinfeld 1978b). Applying this approach to aircraft noise requires the difficult empirical task of accurately measuring the NEF values for each location and obtaining data on other important

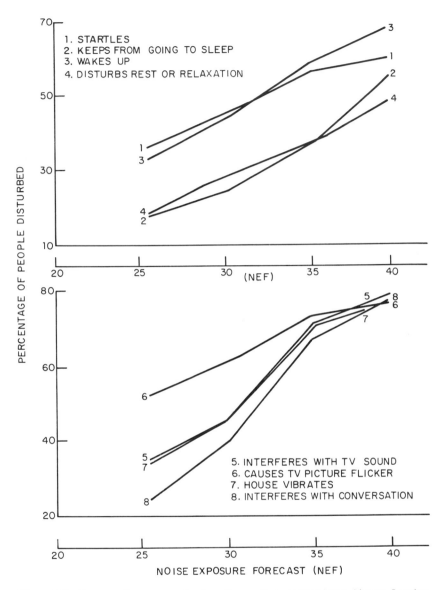

Figure 2.3 Community response to aircraft operations at Heathrow Airport, London.
Source: Federal Aviation Administration 1977, p. 13.

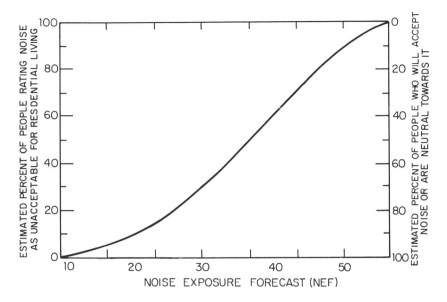

Figure 2.4   Attitudes toward aircraft noise in residential community. Source: Federal Aviation Administration 1977, p. 17.

housing attributes. Omitted information may bias the estimates of the value that people place on noise improvements if the omitted variables and noise are correlated. For example, if noisy neighborhoods also tend to have more industrial development, and industrial development is not included in the relationship, this estimation technique would over-state the willingness to pay for noise reduction by including the distaste for being near industry.

Nelson (1979) summarizes thirteen major empirical studies of aircraft noise and housing prices, standardizing the results by calculating the percentage decrease in housing price per NEF unit, which assumes a linear willingness-to-pay function. The mean noise discount for a change of one NEF unit is 0.62 percent, with noise discounts in the range of 0.4–1.1 percent per NEF unit. In an area with an average housing value of $100,000, households would therefore value a one-unit change in noise level at about $620, with the likely range of $400–$1,100. Because they are based on housing prices, these estimates represent the dis-counted present value of the NEF change; an annual value can be obtained by multiplying by the discount factor. For example, a 10 percent discount rate implies a value of $62 per year for the one-unit change.

None of the existing studies of aircraft-noise valuation employs the conceptually correct two-step procedure to test whether the relationship between NEF level and marginal willingness to pay is nonlinear. Moreover, it is virtually impossible to determine whether the estimates are biased because of omitted variables. Nevertheless, the general agreement of available empirical analyses suggests that bias is relatively small. The assumption of a linear relationship between NEF and willingness to pay over the range of NEF 30 to NEF 45 also appears justifiable from the results of annoyance surveys, which do not indicate any significant nonlinearities in the percentage of persons annoyed.

Options for Reducing Noise Damage

The options to limit exposure to aircraft noise can be put into the following five categories: reductions in the noisiness of aircraft engines, changes in takeoff and landing procedures, changes in the number or the timing of aircraft operations, receptor changes (such as soundproofing) that reduce the noise that actually reaches people's ears, and land-use changes that reduce the number of persons under flight paths. This section briefly summarizes the noise-control options in order to provide a background for the discussions in chapters 3 and 4 of the likely responses of airlines to economic-incentive schemes.

Aircraft engine noise

Propeller aircraft, which make considerably less noise than jets, have become a smaller and smaller fraction of the commercial fleet since jets were introduced in 1958. Because of their greater speed, longer range, and less costly operation, jets rapidly replaced piston and turbo-prop aircraft (Straszheim 1969). By 1978, jets accounted for 88 percent of the roughly 2,500 aircraft in the commercial fleet (Federal Aviation Administration 1978, p. 11).

The jet fleet has undergone dramatic changes. Most early commercial jet planes were powered by pure turbojet engines, but now most commercial jets have the quieter and more powerful turbofans, which obtain thrust from both the primary jet and the fan exhaust. Turbofan engines, however, generate noise from the jet exhaust system (created by the turbulent mixing of high-velocity exhaust gases with the ambient air) and from the fan and compressor system (which emits the particularly annoying high-frequency tones; see U.S. Department of Housing and Urban Development 1972). Three major generations of turbofan jet engines currently dominate the airlines' fleets (U.S. Environmental Protection Agency 1980, pp. 11–13). The four-engine Boeing 707 and Douglas DC-8, powered by first-generation engines of the JT3D type, operate primarily on medium and long routes and carry approximately 150 passengers. The second-generation engine, the JT8D, is used in two- and three-engine aircraft such as the Boeing 727, the Boeing 737, and the Douglas DC-9, which fly on short- to medium-range routes and carry 80–120 passengers. The third type is the high-bypass-ratio engine, which reduces jet exhaust noise by increasing the ratio of the air bypassing the combustion chamber to the air flowing through it. This engine also reduces fan noise, particularly the annoying high-frequency noises. Aircraft developed after 1969 all employ these quieter engines, and some of the older aircraft have now been refitted with them. The Boeing 747, which carries about 365 passengers, has four JT9D engines and is used primarily on long routes. The three-engine Douglas DC-10 and Lockheed L-1011 also employ quiet-fan engines, flying medium-range routes and carrying about 250 passengers.

Several government-funded research projects in the late 1960s and early 1970s investigated the possibility of fitting noisy aircraft engines with noise-reduction equipment. The two major retrofit techniques developed were to add sound-absorption material (SAM) to the engine nacelles and bypass ducts and to reduce jet velocity by using larger-diameter, high-bypass-ratio fans (Refan). Government tests indicated that equipping JT3D and JT8D engines (which together account for about 1,500 aircraft, about 70 percent of the U.S. air-carrier fleet) with SAM could produce significant reductions in noise. Tests on JT8D aircraft showed that Refan could reduce noise levels even more.

The technology for a fourth generation of aircraft engines, more fuel-efficient and less noisy, was developed in the 1970s, but the planes will not be available until the mid-1980s. The major U.S. representatives of this new technology are the Douglas DC-9-80 and the Boeing 757 and 767. Airlines are just now ordering these aircraft in significant numbers. In November 1980, Delta placed a $600 million order with Boeing for 757s—the largest dollar purchase in the history of the aircraft industry (*New York Times*, December 19, 1980, p. A1).

Takeoff and landing procedures

To lessen noise impacts, most airlines reduce engine thrust at takeoff according to the procedures developed by the Air Transport Association, Northwest Airlines, the Airline Pilots Association, and the Federal

Aviation Administration (Miller 1979, p. 68). Although thrust cutback can decrease noise in a highly populated area under the flight path, it also reduces the climb gradient so that the noise downrange increases. In effect, these procedures thus trade some reduction in high-level noise near the airport for an increase in lower-level noise further away. For example, one cutback procedure reduces noise about 2 EPNdB at distances 4–14 miles from the takeoff point and increases noise about 2 EPNdB beyond that distance (U.S. Environmmental Protection Agency 1973).

Landing procedures to reduce noise include flap management, two-segment approaches, steep approaches, and decelerating approaches. An approach made with less than full landing flaps reduces noise by lowering airframe drag and hence requiring less power to keep the aircraft in flight. In general, the lower the altitude when changing from approach to landing flaps, the smaller the area exposed to a given noise level. Data for an L-1011 show that delaying the transition altitude from 600 to 300 meters reduces the area within the 70-dB contour from 14 km^2 to 11 km^2; the area within the 80-dB contour, however, seems relatively insensitive to such changes (U.S. Environmental Protection Agency 1973, p. 2-9). By 1976 all airlines except Pan American used reduced-flap procedures, and in 1977 the FAA promulgated a regulation requiring delayed-landing-flap procedures.

Instrument landing systems at most airports are set to guide aircraft along a 3° glide slope to the runway. A two-segment approach (5°/3° or 6°/3°) or steep approaches (5° or 4°), however, are likely to contribute to noise reduction by creating more distance between the aircraft and the community and by lowering thrust. Comparing noise levels of a 737 using a 5°/3° two-segment approach and 4° and 5° steep approach profiles, a recent report concludes that these landing procedures can reduce noise significantly (Hastings et al. 1977).

There are essentially no extra capital or operating costs associated with thrust-cutback takeoffs and delayed-landing-flap approaches. In fact, Northwest Airlines' takeoff procedure reportedly saved about $3 million on fuel in 1975 (Kozicharow 1976). To implement the two-segment approach procedure, however, requires investments of about $35,000–$45,000 (in 1975 dollars) in avionics for each aircraft, plus about $50,000 in ground equipment for each runway (Meindle et al. 1976, p. 5-2).

Table 2.2 The FAA's categories of air carriers.

	No. of carriers	No. of aircraft
Trunk	11	1,706
Local service	18	491
All cargo	3	36
Supplemental	6	77
Helicopter	1	3
Intrastate	5	60
Contract	18	101
Travel clubs	16	21
Total	78	2,495

Source: U.S. Department of Transportation 1978, p. 21.

Aircraft operations

Noise exposure is affected by overall aviation activity, particularly that
of the air-carrier industry (air carriers, general aviation, air taxis, and
commuter airlines). Defined by the FAA as "any operator of large
aircraft that transports passengers or cargo for hire," air carriers are of
primary concern since they operate the large jets that generate the most
noise. The general-aviation fleet, which accounts for about 80 percent
of aircraft operations, consists mostly of much less noisy single-engine
piston aircraft. Air taxi services and commuter airlines both operate
relatively small planes and account for little activity nationally.

The FAA classifies air carriers into the eight categories listed in
table 2.2. A small number of national (or trunk) airlines and a somewhat
larger number of local-service airlines dominate the air-carrier industry:
Eleven trunk carriers and eighteen local-service carriers account for 88
percent of the industry total. The FAA expects air-carrier revenue passenger miles to increase by 5.3 percent a year between 1978 and 1990
and passenger enplanements to grow by 4.7 percent a year over the
same period (U.S. Department of Transportation 1978, p. 11). Because
of the increasing use of larger planes and the likelihood that load factors
(the percentage of available seats that are occupied) will also rise, however, the number of aircraft operations is only projected to grow by
2.1 percent a year over this period.

One option for lowering noise exposure is to reduce aircraft operations,
or at least to slow their growth. Simply decreasing overall activity,
however, generates only small noise benefits. As table 2.3 demonstrates,
a 50 percent cutback in Logan Airport's flights (applied evenly around

Table 2.3 Relationship between aircraft operations and NEF values for Boston's Logan Airport.

Percent reduction in operations	0	21	37	50	60	68	80	90
Percent operations remaining	100	79	63	50	40	32	20	10
Decibel reduction	0	1	2	3	4	5	7	10
Remaining NEF value	40	39	38	37	36	35	33	30

Source: Massport Master Plan Study Team 1975, p. II–19.

the clock to all types of aircraft) would result in only a 3-dB reduction in noise. This finding results from the calculation of NEF (the measure of noise exposure) as an average of sound energy over a 24-hour period; halving the total number of operations halves the sound energy, producing only a three-unit change in the logarithmic decibel scale.

Reducing night flights or using quieter planes, in contrast, will yield greater noise-exposure improvements. As mentioned above, the NEF calculation adds a penalty for nighttime operations to account for the disturbing of sleep; thus, although night operations represent about 10 percent of total commercial operations at Logan, they are responsible for a much greater part of the airport noise. A 10 percent reduction in total operations applied to all types of aircraft at all times of day would reduce NEF by about 0.5 dB; eliminating all night operations would reduce NEF more than 4 dB (Massport Master Plan Study Team 1975, p. II.20).

Sensitivity to noise
The annoyance that noise causes can also be reduced by changing the circumstances of those exposed to it. Examples of such changes are barrier construction, soundproofing, and the purchase of noise easements.

Trees or solid structures are often useful for screening out highway noise, but they are generally less effective in eliminating aircraft noise. Barriers can reduce noise levels by as much as 10 EPNdB in areas adjacent to the airport, primarily by muffling engine warmup noise (U.S. Department of Housing and Urban Development 1972, p. 84). The Airport Development Aid Program, authorized by the Airport and Airway Development Act of 1976 (Public Law 94-353), permits as allowable costs "the purchase of noise-suppressing equipment, the construction of physical barriers, and landscaping for the purpose of diminishing the effect of aircraft noise on any area adjacent to a public airport. . . ." The amount actually spent has been small. From the enactment of the program to June 1979, only $500,000 was granted for the construction of physical barriers (U.S. Environmental Protection Agency 1980, p. 14).

Soundproofing is the most promising option in this category. Homeowners can soundproof their homes (Zeckhauser and Fisher 1976), and government could subsidize noise insulation of existing units or raise the standards of building codes. A U.S. Department of Housing and Urban Development report (1972) estimates that soundproofing can reduce interior noise by up to 25 EPNdB (depending upon the type of construction and the extent of soundproofing modifications), but concludes that "requiring a high degree of noise insulation in residential construction has limited application as a strategy to reduce the impact of aircraft noise" (p. 213). The major three difficulties cited are that residential soundproofing provides only limited relief because outside noise is not affected, the cost is high, and there are practical difficulties in achieving large noise reductions. The HUD report estimates that soundproofing a 1,200-square-foot house would cost (in 1972 dollars) about $3,600, $5,400, and $9,600 for noise decreases of about 5 EPNdB, 10 EPNdB, and 15 EPNdB, respectively (p. 226). (These soundproofing studies predate recent energy-price increases; U.S. Environmental Protection Agency 1980 indicates that residential soundproofing may be more cost-effective now because it also reduces heating and cooling requirements.)

Airports commonly purchase easements limiting building height from owners of land on approach zones to provide for safe approach. Noise easements are also based upon a principle of acquiring partial interest in land—in this case, the interest in being free from aircraft noise. Homeowners can acquire easements even when they are not offered if they are successful in bringing suit for compensation equal to the noise-related decrease in market value (Baxter and Altree 1972).

The cost of a noise easement obviously varies with the extent of the noise damage. According to U.S. Housing and Urban Development 1972, easement awards are typically 10–20 percent of market value—the same order of magnitude as soundproofing costs, although these estimates are probably for units exposed to the most noise. The drawbacks of easements are that they do not actually reduce noise damage, and they can become outdated if noise increases. To deal with this second objection, Baxter and Altree (1972) propose time-limited easements, to be recalculated every two or three years.

Compensating airport neighbors raises practical as well as equity problems. Though a general constitutional right exists that property should not be taken without just compensation, this principle cannot justify complete compensation of all airport neighbors. Current residents may well have bought noise-affected properties at a discount, and thus the real losers are the previous owners. These earlier residents may be very difficult to trace, and the passage of time may make them less compelling equity targets (Baxter and Altree 1972, pp. 25–28). The question also arises whether owners or renters should be compensated. If rents reflect the disadvantages of noise, presumably only landlords should receive compensation; it is likely, however, that there would be strong pressure to compensate tenants as well. In addition, a compensation scheme may encourage people to move into an area, or at least not to leave, thus providing precisely the wrong incentives to obtain the least costly means of reducing noise damage (Olsen and Zeckhauser 1970; Baumol and Oates 1975).

Even if the right people were compensated, the equities of allocating noise-charge revenues are not clear. One could argue that airport neighbors ought to be spared the intrusion of aircraft noise, but the argument might also be made that freedom from noise intrusions is not an absolute right and that some sharing between airport neighbors and airport users is appropriate (Coase 1960; Baxter and Altree 1972).

Land use

The number of persons overflown can be reduced by relocating the airport or the neighboring population or by restricting population growth in noise-affected areas. Although such land-use changes are often mentioned as important components of an overall noise-abatement program (see, for example, Federal Aviation Administration 1976, p. E-20), their actual importance has been limited. Because of the intense opposition of surrounding communities, of travelers, and of airlines, locating major new airports in outlying areas is not a likely alternative. Milch (1976) reports the blocking of new airport proposals in New York, Miami, Chicago, Minneapolis–St. Paul, and St. Louis, pointing out that only a handful of new airports have been constructed since the advent of the jet age. Washington's Dulles Airport and the Dallas–Fort Worth airport are the two most prominent examples of new facilities designed to reduce noise damage by relocating in undeveloped areas.

Although the record is somewhat difficult to document, it appears that prohibiting growth around airports through zoning or purchase also plays a limited role in reducing noise damage. Blitch (1976) describes the conflict between Oakland International Airport and the city of Alameda, California, over 900 acres of vacant land exposed to very high noise levels; despite the airport's vigorous complaints and the recommendations of several studies to prohibit incompatible land use, the city declined to rezone the area to prohibit residential development. The growth of residences in the flight paths of Dulles and Dallas–Fort Worth also suggests that land-use controls do not prevent widespread noise damage, although they may still deter construction in the most noise-impacted areas around the airport.

Removing people from the noisiest areas is costly unless the land can be redeveloped for commercial or industrial use. Los Angeles International Airport paid $144 million to acquire surrounding land, mostly in the late 1960s (U.S. Environmental Protection Agency 1980, p. 14). Local opposition to an airport's buying up land often arises, because residents fear abandonment of housing and destruction of the neighborhood. For example, Massport (the Massachusetts Port Authority) has offered to buy houses in severely affected areas around Logan Airport, but has acquired few properties despite the generous purchase conditions. Nevertheless, airports have spent $22.6 million in federal grant funds to acquire land for noise-control purposes (U.S. Environmental Protection Agency 1980, p. 14).

Government Controls

As in most regulatory areas, the federal, state, and local governments share responsibility for controlling aircraft noise. However, regulation of aircraft noise involves another major quasigovernmental actor: the airport operator. A complex mix of policy statements, legislation, and court decisions has shaped the division of responsibility (Nelson 1978; Nierenberg 1978). Two major issues that remain controversial are what areas of noise control are preempted by the federal government through its control of interstate commerce and what liabilities exist for aircraft-noise damage. In general, the federal government, through the FAA, has regulated the inherent noisiness of aircraft engines and takeoff and landing procedures, while local airport operators have regulated changes in receptors (soundproofing and so forth) and in land use. Control of aircraft operations is somewhat less clear because of the conflict between local airports' efforts to limit the noisiness or the timing of flights and the federal government's interest in unrestricted interstate commerce.

The implementation of many measures to control noise is circumscribed by other objectives, such as safety or free movement of commerce. These constraints have the effect of limiting the actions airlines can take to reduce noise damage.

Federal regulation
Under the Federal Aviation Act of 1958, the federal government assumed responsibility for regulating airspace use and management, air traffic control, safety, and source noise. The FAA thus controls takeoff and landing procedures (including flight profiles, tracks, and turns) and certifies aircraft for compliance with safety and noise regulations. State and local governments have no legal power to set contrary regulations. The rationale for federal preemption is clear: A safe and reliable air transport system requires consistent standards.

Congress granted the FAA its authority to regulate aircraft noise in the 1968 amendments to the Federal Aviation Act, known as the Noise Control and Abatement Act. The major actions under this mandate have been to control noise "at its source," an emphasis consistent with the FAA's general certification procedures for aircraft safety and with congressional intent. While acknowledging that other approaches to aircraft noise control must be thoroughly studied and employed, the Senate report on the Noise Control and Abatement Act also states that "the first order of business is to stop the escalation of aircraft noise by imposing standards which require the full application of noise reduction technology" (cited in Nelson 1978, p. 46).

The FAA's chief regulations for the control of aircraft noise are known as Part 36 standards. In November 1969, the FAA established maximum noise levels for newly designed aircraft heavier than 75,000 pounds certified after December 1, 1969. Standards were established for takeoff, approach, and sideline noise; aircraft not meeting the standards would be denied a certificate of airworthiness. As figure 2.5 illustrates, the newer aircraft (Lockheed L-1011, Douglas DC-10, and Boeing 747) meet the Part 36 noise standards and the older aircraft do not. In 1973, the FAA extended the Part 36 noise limits to newly manufactured aircraft of pre-1969 design.

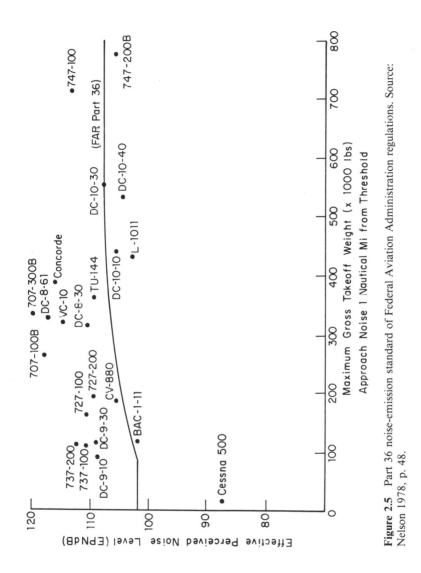

Figure 2.5 Part 36 noise-emission standard of Federal Aviation Administration regulations. Source: Nelson 1978, p. 48.

In 1976, jet aircraft already in service were required to meet the Part 36 standard, either by retrofit or replacement. The phased timetable for compliance was as follows:

• 747s within six years (by 1983), with half to be completed within four years,

• 727s, 737s, DC-9s, and BAC-111s within six years (1983), with half to be completed within four years, and

• 707s, 720s, DC-8s, and CV-990s within eight years (1985), with one-fourth to be completed within four years and half to be completed within six years.

The Part 36 requirement was further amended in March 1977 to require the next generation of commercial aircraft to meet even stricter noise standards.

Extension of the Part 36 standards to the existing fleet—the retrofit rule, as it is often called—has been the most controversial of the FAA's source-noise requirements. Applying the standards for newly manufactured aircraft was much less controversial, largely because manufacturers were already using quieter engines owing to the happy coincidence that designing for better fuel economy also reduces engine noise. Through the Air Transport Association, the airlines have steadfastly opposed the Part 36 extension, arguing that the benefits are not commensurate with costs, that the compliance schedule was impossible to meet (because of the limited availability of retrofit kits and the airlines' limited resources), and that the requirement would compromise fuel efficiency and safety objectives (Federal Aviation Administration 1976). In 1978 and 1979, Congress debated proposals to relax the controls or to fund noise control out of the federal excise tax on airline tickets. On November 20, 1979, the FAA announced that it would extend the timetable for retrofitting the planes having two or three JT8D engines (44 *Federal Register* 12021).

In addition to setting noise standards, the FAA has modified operational procedures to reduce noise. In 1967, for example, the FAA ordered air traffic controllers to clear aircraft for landing only from the highest possible altitudes, a measure designed to keep aircraft as far as possible from populated areas. In 1976 the FAA also required the use of minimum certified landing flaps to reduce noise. A draft advisory circular explaining and encouraging airlines to adopt a less noisy takeoff procedure was issued in 1979. (It is difficult to determine what effect this voluntary program will have on aircraft operations.)

By controlling the airport development program, the FAA can also affect airport noise indirectly. The FAA's large grant-in-aid program, funded by an 8 percent ticket tax for airport development, has since 1970 included environmental considerations among the criteria for project approval. The Airport Act was amended in 1976 to allow the use of program funds for noise-abatement devices and land acquisition. At least one commentator (Nierenberg 1978, p. 172) believes that noise reduction under the airport development program has been minimal.

In the Comprehensive Noise Control Act of 1972, Congress amended the Federal Aviation Act to add "protection of the public health and welfare" to the original statement of purpose. It also gives the Environmental Protection Agency a role in aircraft noise abatement by enabling it to propose rules. Notices of EPA proposals are published in the *Federal Register*; if the FAA does not adopt an EPA proposal after a "reasonable time," it is obliged to publish an explanation there. The FAA, however, retains ultimate responsibility for regulating aircraft noise. Since this somewhat unusual statutory scheme was established, in 1974, the EPA has submitted eleven proposals to the FAA. (See p. 173 of Nierenberg 1978 for a list of the proposals and the FAA's actions.)

State and local control
The Noise Control Act of 1972 states that "primary responsibility for control of noise rests with state and local governments" (cited in Nierenberg 1978, p. 175). Because of the federal government's preemption in many areas, however, the noise-abatement authority of state and local governments is severely constrained by court-imposed limitations on their rights to act under general police powers. Since the federal government has exclusive control over the management of air space, for example, state and local authorities cannot prohibit or regulate overflights. In *City of Burbank* v. *Lockheed Air Terminal* (411 U.S. 624, 1973), this principle of preemption was extended to other restrictions on aircraft operations. The Supreme Court struck down an 11:00 P.M.–7:00 A.M. curfew on jet aircraft imposed by the city of Burbank, indicating that other types of police-power restrictions (such as those on the type of aircraft using a particular airport) are also prohibited. (The Court specifically excluded the airport operator from these prohibitions.)

These legal constraints leave only land-use control as the major tool that state and local governments can use to reduce the impact of aircraft noise. As discussed above, noise exposure can be lowered by prohibiting residential growth near airports, relocating residences from noise-impacted neighborhoods, and locating public facilities such as schools and highways in unaffected areas. It is not clear that these measures have actually been used to reduce noise exposure.

Unlike other states, California has enacted legislation imposing responsibilities for noise abatement on airport proprietors (Nierenberg 1978, p. 176). The California Noise Act of 1969 directs the state Department of Aeronautics (now the Department of Transportation) to adopt ambient-noise standards for airports operating under state permits. State regulations require airports to achieve a gradual reduction in noise exposure that will ensure that by 1985 no area will have a "community noise equivalent level" above 65 (roughly equivalent to NEF 30). Although the regulations originally required establishment of single-event noise-exposure levels, this provision was struck down in *Air Transportation Association of America* v. *Crotti* (389 F. Suppl. 58, N.D. Cal., 1975) as an intrusion into the exclusive domain of federal control of aircraft flights and operations, and thus invalid under *Burbank*. California airport operators have responded to the state regulations by implementing a variety of control measures. For example, San Diego has a midnight–6:00 A.M. curfew on takeoffs and landings by aircraft not meeting Part 36 standards. However, other operator responses to the California directive have become embroiled in controversy.

Airport operators
The role of airport operators in controlling aircraft noise is rather unclear, and court cases and federal and state policy statements have contributed to the confusion. The basic dilemma is that the airport operators' liability for noise damage may exceed their authority to control its causes.

In *Griggs* v. *Allegheny County* (369 U.S. 84, 1962), a Pittsburgh resident sued the Greater Pittsburgh Airport, the airlines, and the United States government for damage suffered from aircraft noise. The Supreme Court found the airport proprietor alone liable, reasoning that the proprietor had planned the airport's location, the configuration of the runways, and the aircraft operations and thus should bear the ultimate burden. The liability for damages is based on the constitutional requirements that just compensation be given for property taken for a public purpose.

The courts and the federal govenment have been much less clear about the actual authority of airport proprietors to control aircraft noise. In its *Aviation Noise Abatement Policy*, the FAA states the following:

The power thus left to the proprietor—to control what types of aircraft use its airports, to impose curfews or other use restrictions, and, subject to FAA approval, to regulate runway use and flight paths—is not un limited. Though not preempted, the proprietor is subject to two important constitutional restrictions. He first may not take any action that imposes an undue burden on interstate or foreign commerce and, second, may not unjustly discriminate between different categories of airport users. (U.S. Department of Transportation 1976, p. 34)

As discussed above, the 1973 *Burbank* case did not include airport proprietors in its prohibition against several noise-control actions by local governments acting under their police poweis. (The distinction between local governments and airport proprietors is often blurry, since local governments usually own and operate airports.) But *Burbank* also did not delineate what local governments could do as airport proprietors. In *Crotti*, the Northern California District Court held that requiring airports to achieve reductions in noise exposure using a variety of land-use and operational changes was not invalid *per se*. The court stated the following:

It is now firmly established that the airport proprietor is responsible for the consequences which attend his operation of a public airport; his right to control the use of the airport is a necessary concomitant, whether it be directly by state police power or by his own initiative. The correlating right of proprietorship control . . . necessarily includes the basic right to determine the type of air service a given airport proprietor wants its facilities to provide, as well as the type of aircraft to utilize those facilities. (quoted in Nierenberg 1978, p. 176)

In *National Aviation* v. *City of Haywood* (418 F. Supp. 417, N.D. Cal., 1976), the same district court upheld the constitutionality of the airport proprietor's right to prohibit all aircraft exceeding a certain noise level from landing or taking off from a city-owned airport. The court did indicate, however, that the FAA had the authority to preempt such regulation.

Other court decisions suggest a more circumscribed role for airport proprietors. Much of the controversy has arisen from efforts of the state of California to require controls. In 1978, the California Department of Transportation granted a variance from other requirements to the San Diego Unified Port District on the condition that the district extend the hours of its curfew on aircraft operations at Lindbergh Field.

San Diego is now challenging the constitutionality of this condition. In October 1978, the U.S. District Court for Southern California granted a preliminary injunction enjoining the California Department of Transportation from terminating the variance (*San Diego Unified Port Authority* v. *Gianturco*, 12 ERC 1046, 1978), suggesting that the curfew extension would not be upheld.

Uncertainty over the legality of airport operators' actions is exacerbated by the potential conflicts between the objectives of controlling noise and increasing airline competition. By eliminating Civil Aeronautics Board control over routes, fares, and services, the Airline Deregulation Act of 1978 was designed to foster competition and thus lead to lower fares and better services (Meyer et al. 1981). Without route restrictions, California airports (among others) experienced substantial increases in traffic, which increased congestion and noise. Proposals to deal with these new conditions illustrate the conflict between goals. For example, the San Francisco Airports Commission adopted a resolution in January 1979 that would require air carriers serving the area for the first time to use only Part 36 aircraft. In San Diego's noise-control plan, the airport operator proposed a moratorium on new users in order to evaluate the possible consequences of airline deregulation on noise control. Operators of the Burbank Airport also proposed a rule that would authorize additional air-carrier operations only after determining that they would not increase the noise-impacted area around the airport. All three of these noise-abatement proposals constrain new entrants into the lucrative California markets. Commenting on the San Francisco proposal, the general counsel for the Civil Aeronautics Board stated the following:

The action of the Commission conflicts sharply with federal policy as set forth in the Airline Deregulation Act of 1978 (PL 95-504), which was enacted to foster entry by air carriers into new interstate markets. In addition, the resolution's applicability to new but not incumbent airlines appears to violate the Constitutional prohibitions against discriminatory local regulations to interstate commerce. . . .

Finally, any attempt to dictate the type of aircraft to be operated — even if applied even-handedly — may conflict with a detailed federal noise policy developed by the Department of Transportation. . . . The federal government has a comprehensive noise policy that phases out noisy aircraft over a four-year period and requires airlines to begin flying quieter aircraft in 1981 and to conform their entire fleets by 1985. If the San Francisco airport, as it purports to do, or any other airport enforces more stringent equipment requirements, there would be a conflict with an area of regulation that the federal government may have preempted and there would probably be a serious disruption in interstate air service because planes lawfully departing from, say, Chicago, New York, or Denver would not be able to land in San Francisco (Civil Aeronautics Board, February 13, 1979).

Until the federal government or the courts clarify the bounds on the airport proprietor's control over operations, the legality of actions such as the San Francisco resolution will continue to be questionable.

Chapter 3

Federal Noise-Control Strategies

With this chapter begins the detailed evalution of noise-control strategies. The focus here is on alternative strategies the federal government could adopt and implement. The analysis emphasizes the strengths and weaknesses of charges versus command-and-control regulations rather than any particular policy.

To provide a detailed comparison, four national noise-control policies have been selected for discussion: extension of Part 36 noise standards to the existing fleet (the retrofit rule), a uniform national noise charge levied on all aircraft operations, a retrofit rule that varies by airport, and a noise charge that varies by airport. Analysis of these four policies helps address the advantages and disadvantages of substituting an economic-incentive scheme for traditional command-and-control regulation and the gains to be achieved by tailoring the severity of any policy—charges or regulations—to the severity of the environmental problem.

Because of its importance in federal noise-control efforts and the availability of estimates of its costs and benefits, the retrofit rule is chosen to represent the standards approach. The principal economic-incentive scheme evaluated in this chapter is the noise charge, although some analysis of the noise-rights scheme is also presented.

The chapter first discusses the calculation of an actual charge scheme. One of the major arguments against econonomic-incentive schemes is that there is not enough information to specify the ideal charge that will equate marginal benefits and marginal costs. An actual proposal would require updated date and other refinements, but it is possible to calculate a charge for airport noise that corresponds closely to the theoretical ideal.

Since the major advantage of economic-incentive schemes is that they are more efficient than equivalent regulations, the first comparisons center on the costs and benefits of the regulatory alternatives, beginning with the retrofit rule. Because they depend upon detailed operating characteristics of the airline's fleets, the precise benefits and costs under the noise-charge schemes (even the national uniform strategy) are impossible to estimate. The method for comparing the net benefits of these other schemes, therefore, is to use the retrofit rule as a baseline and to analyze the likely improvements in cost-effectiveness under the charge scheme.

Many public-policy comparisons concentrate on the overall costs and benefits of alternatives, a limitation that may be appropriate when the choices are reasonably narrow. To compare broad regulatory approaches, however, this chapter includes three additional criteria: administrative and enforcement difficulties, flexibility to deal with changes over time, and incentives to develop new technology. The existing literature has addressed all three factors in general terms (Baumol and Oates 1979); the contribution of this study is that it provides some sense of their importance in the context of aircraft noise control. The final two issues the chapter addresses are the differential impact on various airlines and the use made of charge revenue—factors that may influence adoption of the noise-charge approach.

Specifying Noise Charges

Setting an ideal charge requires more information on the costs and benefits of control than is usually available. To determine a price that equates marginal benefits and marginal costs, the regulator must estimate the schedules relating marginal costs and benefits to the level of control. The difficulty of making such estimates, together with reluctance to set environmental goals solely on the basis of efficiency, probably explains why charge proposals do not seek to achieve the most efficient level of control, but only to enforce a somewhat arbitrary level of control at lower cost. For example, Nelson (1978, pp. 204–208) calculates a noise-charge scheme based solely on the costs of retrofit for five major aircraft types. Table 3.1 suggests that a noise charge of $25 per EPNdB in 1975 would induce the retrofitting of most older aircraft. (The figure would be $45 in 1982 dollars.)

This cost-based strategy is similar to two other proposed charge schemes: the noncompliance penalty system developed in Connecticut and adopted in the 1977 federal Clean Air Act amendments and the "safety-valve" scheme proposed by several economists (Weitzman 1974; Roberts and Spence 1976; Spence and Weitzman 1978). The noncompliance penalty is a fine equal to the money the company saves by not complying with a given regulation (see Drayton 1978), which in the

Table 3.1 Information used to set a cost-based noise charge.

Aircraft	Engine	Number of retrofit candidates	Present value of retrofit per aircraft[a]	Annual cost per landing[b]	Implicit cost per EPNdB reduced[c]
707	SAM-JT3D	217	1.417	231	18.05
DC-8	SAM-JT3D	129	1.249	204	22.67
727	SAM-JT8D	683	0.212	35	6.25
737	SAM-JT8D	156	0.255	42	6.18
DC-9	SAM-JT8D	345	0.212	35	4.43

Source: Nelson 1978, p. 207.
a. In millions of 1975 dollars.
b. Based on a 10-year economic life, 1,000 landings per year, and a 10 percent discount rate. In 1975 dollars.
c. In 1975 dollars.

case of aircraft noise might correspond to the estimates in table 3.1. Under the safety-valve scheme, the government would allow an airline to pay a large charge rather than comply with the standard in every case, thereby setting an upper limit on noise-control expenditures. Although these charge schemes introduce economic flexibility, they are primarily intended as enforcement mechanisms rather than approximations of an ideal charge.

As discussed in chapter 2, the one situation where an ideal charge can be set without estimating the complete schedule of costs and benefits is if the relationship between households' valuations of the benefits of increasing controls is known (or can be estimated with some confidence) and contains no major thresholds. When these conditions are met, the charge can equal marginal benefits and the polluters' reactions will determine the efficient level of control. Airport noise is one such case. The large body of housing-price evidence provides an ideal means of estimating people's valuations of greater quiet, since aircraft noise is easily recognizable and highly localized. These studies also suggest that the marginal benefit of reducing NEF level is approximately constant over the relevant range; there is no major threshold where noise becomes intolerable. Placing a dollar value on noise-control benefits also does not entail the complexities of distinguishing the contribution of a pollutant to higher death or illness rates, or of valuing lives saved.

Calculating a uniform noise charge on emissions
A study by the Council on Wage and Price Stability (COWPS) estimated that national aircraft damages in 1975 were $3.25 billion (in 1975

dollars), on the basis of Nelson's (1978) calculation of the value that households place on noise above NEF 30 and the U.S. Department of Transportation's 1974 estimates of the number of persons exposed to various NEF levels (see chapter 2). Assuming a 10 percent discount rate, damages would thus be $325 million per year. Assuming a threshold level of 98 EPNdB (corresponding to NEF 30), and on the basis of data on the total number of aircraft operations by aircraft type and their noise levels (U.S. Department of Transportation 1976, pp. 36–38; Council on Wage and Price Stability 1977, p. 21), the total number of EPNdBs emitted above the threshold in 1975 was approximately 45 million. Dividing national damages by national emissions generates a national noise charge of $7 per EPNdB. (This noise charge would be $12 per EPNdB in 1982 dollars.)

The $7-per-EPNdB estimate, of course, is only an illustration. The purpose of this calculation is simply to demonstrate that a plausible noise charge can be set. To refine this estimate might require additional data on noise exposure or more detailed property-value studies. The average charge, however, is not likely to change dramatically; using Nelson's lower estimate of the value that households place on noise exposure, for example, implies that the average charge drops from $7 to $5 per EPNdB.

The noise-charge formula
The major drawback of the national uniform charge is clear: Not all EPNdBs cause the same annoyance. The largest source of variation is probably airport location; a given aircraft operation at LaGuardia Airport will affect a good many more people than the same operation at Dallas–Fort Worth. In addition, a night flight is more annoying than a day flight. Annoyance surveys also show that the change in annoyance increases with EPNdB (that is, the marginal is greater at higher EPNDB levels). The uniform noise charge or standard ignores these considerations and would result in airlines' equalizing the marginal cost of reducing noise emissions per decibel. The appropriate objective, however, is to equalize the marginal cost of removing one unit of noise exposure.

The best basis for measuring noise exposure, and therefore for setting the noise charge, is NEF levels. Airlines would be charged for all NEF-households their flights cause. For example, if an airline's flights at Logan increased NEF by an average of 0.5 NEF for 1,000 households, the airline would be charged for 500 NEF-households. But since the

NEF calculation is so detailed, it is not feasible to charge airlines in this way. It is possible, though, to devise a formula that incorporates the major variations in noise-exposure damage—and still preserves the simplicity of the single charge—by assigning weights to different EPNdB values. Since greater noise-exposure damage occurs at night, when the noise is in the higher EPNdB ranges, and when the aircraft overflies populated areas, the noise charge for an individual operation would have the form

$$C = F(\text{EPNdB}, t, A)$$

where C is the noise charge, t the time of day, and A the airport.

The schedule proposed by the Council on Wage and Price Stability in 1977, for example, is based on a doubling of the area exposed to a given noise level with every increase of 5 EPNdB, a twelvefold penalty for nighttime operations (10:00 P.M.–7:00 A.M.), and separate calculations of the noise damage at each airport. The charges in table 3.2 show the COWPS proposal for Boston's Logan Airport, along with the marginal value of each EPNdB; the average COWPS charge per operation is then presented in table 3.3 for each of the 23 major airports. As its originators acknowledge, the COWPS schedule has its limitations. The large jump from 97 to 98 EPNdB, for example, is not consistent with the rest of the charges. The penalty for night flights may be also too crude, because an evening period is not included. Nevertheless, the proposal provides a good illustration of the sort of charge scheme that could be developed.

Marketable-rights strategy

The case for noise charges implies that the marketable-rights strategy is less attractive on *a priori* grounds: If the benefits from noise reduction are approximately the same for each household-NEF, it makes more sense to fix prices than to fix the quantity of noise exposure. Determining the reduction in noise exposure at which marginal costs equals marginal benefits would require the government to collect and assess information on control costs. It is useful, nevertheless, to describe a marketable-rights scheme to illustrate how such a market for noise might be defined. (See Nelson 1978, pp. 214–219, for additional discussion.)

Setting a maximum number of decibels permitted, for example, could create a market for aircraft noise. A marketable-rights strategy equivalent to the Part 36 extension would thus limit EPNdBs to the amount generated if all aircraft achieved the Part 36 standards. A single market for EPNdB would, of course, entail the same problems as a single price.

Table 3.2 Example of a benefit-based noise-charge formula applied to Logan Airport.

EPNdB	COWPS charge ($/Operation)		Marginal charge ($/EPNdB)	
	Day	Night	Day	Night
97	0	0	0	0
98	31	374	31	374
99	36	429	5	55
100	41	493	5	64
101	47	566	6	73
102	54	650	7	84
103	62	747	8	97
104	72	859	10	112
105	82	986	10	127
106	94	1132	12	146
107	108	1301	14	169
108	125	1494	17	193
109	143	1716	18	222
110	164	1971	21	255
111	189	2265	25	294
112	217	2601	28	336
113	249	2988	32	387
114	286	3432	37	444
115	329	3943	43	511
116	377	4529	48	586
117	434	5202	57	673

Source: Council on Wage and Price Stability 1977, p. 22.

Table 3.3   Average noise charges for 23 major airports.

Atlanta	13.04
Buffalo	36.55
Cleveland	63.29
Denver	52.11
Dulles	5.64
J.F.K.	113.07
LaGuardia	196.67
Logan	120.18
Los Angeles	53.54
Miami	70.55
Midway	45.57
Minneapolis–St. Paul	43.60
Newark	137.99
New Orleans	15.39
O'Hare	73.29
Philadelphia	25.96
Phoenix	13.04
Portland	0.82
San Diego	102.86
San Francisco	64.49
Seattle	28.37
St. Louis	29.45
Washington National	6.06

Source: Council on Wage and Price Stability 1977, p. 24.

To weight different EPNdBs, however, the total number of marketable rights (MR) could be based upon the formula

$$MR = \sum_i \sum_j \sum_t b_{ijt} (EPNdB_{ijt})$$

where b is the weighting factor and i, j, and t refer to EPNdB level, airport, and time of day, respectively.

Costs and Benefits of the Retrofit Rule

Two major studies have evaluated the retrofit rule: a 1974 report by the U.S. Department of Transportation and a cost-benefit analysis contained in FAA's 1976 environmental impact statement. Nelson's (1978) study reconciling the two studies and summarizing the results is the basis of the following cost and benefit estimates.[1]

Costs

The FAA and DOT cost-benefit studies calculated program costs by specifying a baseline number of noncomplying aircraft in the airlines' fleets, determining a retrofit schedule, estimating the investment and operating costs of retrofitting, and discounting the costs to yield a present value.

Baseline case   Approximately 1,600 aircraft currently operated by principal U.S. air carriers do not meet Part 36 standards and are therefore candidates for retrofitting. A comparison of the number and percentage of noncomplying aircraft assumed in the two studies, presented in table 3.4, indicates that the FAA analysis is somewhat more optimistic about the retirement rate of noncomplying aircraft but less optimistic about the growth in the overall fleet.

Retrofit time schedule   The retrofit rule requires that all aircraft comply with Part 36 standards by the end of 1984. Whereas the FAA cost-benefit analysis assumes that 97 percent of the retrofit modifications would be carried out over the six-year period 1979–1984, the DOT study assumes that retrofit would occur in the four-year period 1975–1978. To provide comparable results, Nelson assumed that the DOT schedule would apply to the 1979–1982 period. The difference in baseline aircraft and retirement schedules in the two studies thus generates a difference in the total number of aircraft requiring retrofitting;

Table 3.4   Projected U.S. air-carrier fleet, as of December 31 each year.

	1972	1976	1978	1981	1987	1991
Federal Aviation Administration assumptions						
Noncomplying	—	1,584 (73%)	—	1,375 (53%)	1,085 (33%)	322 (8%)
Complying	—	583	—	1,238	2,029	3,257
New technology	—	0	—	0	165	574
Total		2,167		2,673	3,279	4,153
Department of Transportation assumptions						
Noncomplying	1,761 (68%)	—	1,588 (56%)	1,512 (47%)	922 (21%)	—
Complying and new technology	829	—	1,248	1,705	3,469	—
Total	2,590		2,836	3,217	4,391	

Source: Nelson 1978, p. 147.

Table 3.5   Noise retrofit investment and lost-time costs per aircraft (millions of 1975 dollars).

Aircraft type[a]	FAA estimate	DOT estimate	Lost-time cost
707	1.200	1.200	0.094
DC-8	1.200	1.020	0.102
727	0.225	0.225	—
737	0.270	0.264	—
DC-9	0.270	0.231	—
747	0.250	—	—

Source: Nelson, 1978.
a. All with sound-absorbing materials.

the FAA study calls for modification of 1,217 aircraft, versus 1,530 in the DOT study (Nelson 1978, p. 149).

Retrofit costs   As the data included in table 3.5 indicate, the DOT and FAA studies provide very similar estimates of the investment costs of retrofitting. The capital cost of installing sound-absorption material is considerably higher for the four-engine 707s and DC-8s than for the two- and three-engine aircraft. Retrofitting the four-engine aircraft imposes additional costs because of the eight-day loss of service during installation; the loss is put at $94,000 and $102,000 per plane for the 707 and the DC-8, respectively.

Since he was unable to determine the basis for the FAA's calculations, Nelson relied on the DOT estimates for operating-cost penalties. The

DOT study assumes that unit operating costs increase by 0.1–0.2 percent for the two- and three-engine aircraft and 0.5–0.6 percent for the four-engine aircraft. In addition, the DOT study indicates that applying sound-absorption material to four-engine aircraft generates a fuel penalty of 0.2 percent.

Total costs of the retrofit program   The program is clearly an expensive one: $771.9 million according to the FAA study and $888.9 million according to DOT estimates. In 1982 dollars, the totals are $1.36 billion and $1.56 billion, respectively.

Benefits

Although the FAA and DOT studies both estimate the benefits of the retrofit program in terms of NEF, they measure benefits differently. The DOT study predicts changes in the population exposed to NEF 30, NEF 40, and NEF 45, whereas the FAA study expresses benefits as a reduction in population within the NEF 30 contour. Only the FAA study attempts to assign a monetary value to noise-abatement benefits: $400 per year (1975 dollars) for each person removed from the NEF 30 contour.

Using the DOT and FAA data on program-induced changes in noise exposure, Nelson also calculated benefits of the retrofit rule on the basis of his own estimates of their dollar value. Nelson assumed that households value a one-unit change in NEF at between 0.7 percent and 1.0 percent of their housing value (Nelson 1978, p. 159). Using an average U.S. property value of $25,122 (1975 dollars) and a 10 percent discount rate, these estimates imply a value between $17.58 and $25.12. Between 1975 and 1987, Nelson assumes that the value of a one-unit change in NEF would grow at a rate of 3 percent per year, and 2 percent per year from 1988 to 2001.

Comparisons of costs and benefits

Alternative estimates of total discounted costs and benefits in 1982 dollars, presented in table 3.6, assume a 4 percent and an 8 percent discount rate, as well as high and low benefit values. In general, these results suggest that the costs of the retrofit plan exceed the benefits; net benefits are positive only under the DOT assumptions and in the high-benefit-value case. Extending Part 36 standards to the existing fleet is thus of questionable value.[2]

Table 3.6 Total discounted costs and benefits of extending Part 36 to noncomplying aircraft, 1977–2001 (millions of 1982 dollars).

Discount rate	Costs	Benefits		Net benefits (costs)	
		Low	High	Low	High
FAA					
4%	1070.8	562.4	803.4	(508.4)	(267.4)
8%	865.5	389.6	556.6	(475.9)	(308.9)
DOT					
4%	1277.0	1131.3	1616.0	(145.7)	339.0
8%	1060.0	922.0	1174.3	(237.9)	(114.4)

Source: Nelson 1978, p. 161: converted to February 1982 dollars.

Costs and Benefits of the Uniform Noise Charge

The negative net benefits of the retrofit rule result from two major limitations of the standards approach: The program requires all aircraft to comply, when compliance for some planes may be very costly and may generate very small noise benefits; and it concentrates on quieting planes to Part 36 noise levels, ignoring the possibilities of achieving more cost-effective reductions by other means. Setting a uniform national noise charge, such as the $7 per EPNdB calculated above, relaxes both these limitations. The question is how important the advantages of this approach are likely to be.

Variations in costs of retrofitting

Requiring all planes to meet a single standard virtually guarantees that marginal costs will not be equal across sources, and therefore that allowing the high-cost sources to pay the charge will increase cost-effectiveness. The advantages of the noise-charge scheme, of course, depend upon the variation in the marginal cost of retrofitting per EPNdB among the pre-1972 aircraft.[3] Under average conditions, the cost differs among the five major aircraft types by a factor of almost 6, ranging from approximately $4 per EPNdB for the DC-9 to $23 per EPNdB for the DC-8 (table 3.1). Costs for the JT8D-powered aircraft are about $6 per EPNdB, compared with about $20 for the JT3D-powered planes. If all planes were flown under average conditions, the standards approach could thus easily duplicate the cost-effectiveness advantages of a $7 charge; one would simply limit the Part 36 requirement to JT8D-powered aircraft.

Since a large part of retrofit costs are fixed, however, the cost per EPNdB depends crucially upon the number of flights over which the

Table 3.7 Annualized costs of retrofit per EPNdB reduced (1975 dollars).

Operations per year	Years remaining before aircraft retirement				
	5	7	10	12	15
DC-8					
1000	34	26	21	19	17
2000	17	13	10	9	8
3000	11	9	7	6	6
4000	8	7	5	5	4
DC-9					
1000	13	10	8	7	6
3000	4	3	3	2	2
5000	3	2	2	1	1
7000	2	1	1	1	1

Source: Table 3.1. To obtain cost per EPNdB reduced, the retrofit cost per operation was divided by the average of the takeoff-noise and landing-noise reductions.

costs are allocated. Using alternative assumptions about the aircraft's remaining life and number of operations per year, table 3.7 shows the variation in cost for a DC-8 (JT3D) and a DC-9 (JT8D). Costs range from $1 per EPNdB for a DC-9 making 7,000 operations per year with 15 remaining years to $34 per EPNdB for a DC-8 making 1,000 operations per year and having 5 remaining years. The solid black line shows the division between aircraft that would be retrofitted and aircraft that would pay a charge of $7 per EPNdB.

The timetable provided under the Part 36 extension regulations may approximate the flexibility that the charge scheme offers. For example, the regulations require DC-8s to be in compliance within 8 years, with one-fourth to be completed within 4 years and half to be completed within 6 years. This phased retrofit should allow aircraft with short lives to be retired or replaced. However, since the percentage requirements in the regulations apply to individual airlines rather than the entire fleet, there is no guarantee that the phasing will prevent the costly retrofitting of planes with short remaining lives.

Options for reducing noise under the uniform charge

A noise charge can, in theory, increase cost-effectiveness by permitting airlines to choose options other than retrofit or replacement. As discussed above, alternative means of reducing noise damages include other controls on aircraft engine noise, changes in takeoff and landing procedures, modifications in the number or timing of aircraft operations, and changes in land-use patterns. Airlines should, according to this argument, be able to judge the costs of these modifications and compare them with savings in noise charges. The cost-cutting climate fostered by airline deregulation should induce airlines to reduce the noise charges they pay.

In practice, improvements in cost-effectiveness depend on what additional options are actually available to airlines under the national uniform charge. These are limited for four reasons. Some options cannot be measured without the monitoring of individual flights. Even with monitoring, other alternatives rearrange noise from one place or time to another; an airline's liability under a national charge would therefore not be reduced and there would be no incentive to make the change. Still other strategies may reduce an airline's overall noise levels (with or without monitoring), but the airline cannot affect the change on its own. Finally, some options are already available under the standards approach.

The major noise-control options discussed in chapter 2 are listed in table 3.8, with an assessment of their limitations in the four categories. The options to control aircraft engine noise do not require monitoring of individual flights, but most are also encourged under the noise-standard approach.[4] Although often referred to as the "retrofit rule," the Part 36 extension also permits airlines to meet standards by replacing planes. The charge scheme does, however, provide incentives for airlines to retrofit planes below the Part 36 level. The prime candidates for further noise reduction are the JT8D-powered aircraft (727s, 737s, and DC-9s), which achieve the Part 36 standard at relatively low cost, but studies of the refanning option for quieting engines below Part 36 levels indicate that reducing noise below the Part 36 standards is extremely expensive. Nelson (1978) estimates that the cost per EPNdB of refanning is $215 for a 727, $251 for a 737, and $180 for a DC-9—far greater than the cost of retrofitting a 707 or a DC-8 (about $25 per EPNdB). Although refanning may be used for aircraft with long lives that make many flights per year, it does little to improve the overall cost-effectiveness of noise control.

Registering noise improvements due to takeoff and landing procedures requires at least some monitoring of individual operations, either the measuring of actual noise or spot-checking for compliance with noise-reducing procedures. Spot-checks would be appropriate if, for example, the FAA determined EPNdB values for two-segment landing and required airlines to submit records on whether that procedure was used

Table 3.8 Airline options under a uniform national noise charge.

	Will reduce charges		Airlines can perform on their own	Option under part 36 standards
	No monitoring	Monitoring		
Aircraft engine noise				
Retrofit to Part 36	yes	yes	yes	yes
Retrofit below Part 36	yes	yes	yes	no
Replacement	yes	yes	yes	no
Development of quieter planes	yes	yes	yes	no
Takeoff and landing				
Preferential runway	no	no	no	yes
Two-segment landing	no	yes	no	yes
Change in flight path	no	no	no	yes
Change in thrust	no	yes	no	yes
Aircraft operations				
Decrease overall operations	yes	yes	yes	no
Shift operations to day	no	no	yes	no
Shift operations to less noisy airports	no	no	yes	no
Use quiet planes at noisy airports	no	no	yes	no
Sensitivity to noise				
Soundproofing and codes	yes	yes	no	yes
Barriers	yes	yes	no	yes
Compensation and easements	yes	yes	no	yes

Land use

Relocate airport or runway	yes	yes	no	yes
Remove houses or prohibit residential growth	yes	yes	no	yes

for each flight. Some changes in takeoff and landing procedures would only redistribute rather than reduce overall noise, and therefore would not be options under a national uniform noise charge.

Even if monitoring is available, modifying takeoff and landing procedures may not be a viable option for the airlines. Pilot organizations and airline industry spokesmen have vehemently opposed certain procedures as compromising flight safety (see, for example, U.S. Congress 1977). Moreover, the airlines cannot adopt the two-segment landing procedure on their own, because it requires additional tower and runway equipment and because the FAA must approve the change. Rather than being areas in which individual airlines can act independently according to their particular circumstances, takeoff and landing procedures are operations where uniformity is almost mandatory. If quieter landing procedures such as the two-segment approach are found to be desirable (because of their noise advantages, costs, and safety), the FAA will likely mandate the change.

The only operational change that reduces payments under a national uniform charge is an overall decrease in activity. Shifts in operations from night to day or from noisy to less noisy airports will decrease overall noise damages and thus may eventually reduce noise charges, but they do not immediately lower the charge any one airline pays. Using quieter planes at noisy airports has the same result. Even the decrease in overall operations is not really an additional option to the airlines, however, because the measure is roughly equivalent to retiring aircraft (another way of meeting the Part 36 standard) if aircraft load factors (passengers as a percentage of capacity) and utilization do not change.

The last two categories of noise-reducing measures in table 3.8, sensitivity to noise and changes in land use, could in theory be implemented if the airlines acted together to reduce the noise charge. Airlines could pay for soundproofing or the removal of houses from noise-impacted areas and get their noise charges lowered to reflect the reduction in damage from overflights. In practice, though, it is unlikely that airlines would undertake the actions as a group. Although airlines could exert pressure on airport authorities or local governments to reduce the incompatibility between aircraft operations and residential quiet by changing building codes or zoning regulations, there is little evidence that such efforts are successful (Blitch 1976). A noise charge may provide airlines with much greater incentives to make such efforts jointly, but a national noise charge is so tenuously connected to changes in noise levels at individual airports that the obstacles to group activity may remain insurmountable.

Summary

In the most advantageous case, the noise-charge strategy would operate if large differences in the cost per EPNdB reduced exist among the non–Part 36 aircraft and if many more cost-effective options for controlling aircraft noise than retrofitting engines are available to the airlines. The worst case would be based on the opposite assumptions: that small differences exist in the cost per EPNdB reduced and that few alternatives are available to improve the cost-effectiveness of control.

The analysis in this section suggests that the aircraft-noise case lies somewhere between these extremes. Although a uniform noise charge would increase the net benefits of control, the size of the improvement would probably be modest. The first promising condition does apply: The cost per EPNdB does vary substantially for different aircraft. However, the advantages of the noise charge over the retrofit rule are small, because the timetable for compliance allows airlines to avoid retrofitting high-cost aircraft.

The second inadequacy of standards—the concentration on source control—does not turn out to be important in the noise case. The airlines have few practical options to reduce noise at lower cost than under the retrofit rule. These options must reduce an airline's noise-charge payments (preferably without requiring monitoring of individual flights), must be possible for airlines to undertake individually, and must not be offered by the standards approach. Only retrofitting below the Part 36 level fulfills all three requirements, and that option is less cost-effective than the retrofit rule. The major constraint is the second criterion. The need to trade off safety and other objectives, or to gain agreement with the government or with other airlines, limits many potentially cost-effective options. Adopting a national noise charge is therefore unlikely to decrease the costs of controlling aircraft noise.

Costs and Benefits of the Noise-Charge Formula

The noise-charge formula is a much more promising means of increasing the net benefits of aircraft noise control than the uniform charge. For one thing, the noise-charge formula offers airlines incentives to redistribute noise to decrease annoyance levels. The large penalties for noise at night and in densely populated areas should induce airlines to shift flights to the daytime, operate quieter planes, and use other airports. For example, one would expect the airlines to make greater use of airports such as Dulles and less of close-in facilities such as Washington National. Since the noise-charge formula entails greater penalties for higher EPNdB, airlines have an incentive to eliminate very noisy planes.

The noise-charge formula may also encourage airlines to work together to reduce noise sensitivity and incompatible land use around major airports. Companies that serve Logan Airport, for example, might agree to soundproof buildings or relocate nearby residents to reduce their noise charges. If residents were allowed to sell a transferable noise easement (an agreement to accept the noise in return for compensation), the airlines could also reduce their charges by making direct payments to airport neighbors. Some potentially cost-effective options, however, are still likely to be unavailable. In particular, takeoff and landing procedures must remain uniform because of their effect on safety.

Although it is impossible to predict how airlines would react to the noise-charge formula, and thus impossible to estimate its precise costs and benefits, we can speculate on the scheme's advantages in two ways. The following section first considers how changing to the noise-charge formula affects retrofit decisions. It then provides estimates of the net benefits of geographically varying standards and compares them with the effects of a geographicaly varying charge scheme.

Which aircraft to retrofit

Table 3.9 compares the annual costs of retrofitting aircraft with various lives with the savings in noise charges, according to the fee schedule effective at Logan Airport. The figures suggest that the problem facing the airlines is much more complicated under the noise formula than the simple noise charge. Deciding whether or not to retrofit their planes or pay the charge now depends on details of each plane's use. For example, retrofitting a 727 making as many as 1,000 daytime flights would not be advisable even if the charge schedule at every airport were as high as Logan's; with 25 percent of the operations at night, however, retrofitting is an attractive option for aircraft making as few as 400 flights. The benefits to an airline of retrofitting a DC-8 also depend upon the cost of the retrofit kit, which will vary with the number of kits sold.

This complexity of the retrofit decision under the noise-charge formula is what makes the scheme economically attractive. Determining which planes fly at night or use airports in densely populated areas, and which

Table 3.9   Comparison of costs and savings in noise charge from retrofitting.

Annualized costs of retrofit (excluding fuel or operating-cost penalties

	Retrofit cost[a]	Estimated annual cost over aircraft life (from time of retrofit)[a]		
		5 yr[b]	10 yr[b]	15 yr[b]
707, DC-8	1,600–3,500	422–923	260–570	210–460
727	300	79	49	39
737	360	95	59	47
DC-9	360	95	59	47
747-100	330	87	54	43

Annual savings in noise charges with retrofit

	Total operations	Savings[a]	
		100% Daytime operations	75% Daytime operations
727-200	200	7.9	29.4
	400	15.8	58.8
	600	23.7	88.2
	800	31.6	117
	1000	39.5	147
DC-8-61	200	44.8	183
	400	89.6	336
	600	134	549
	800	179	732
	1000	224	916

Source: Miller 1979, pp. 130, 132.
a. Thousands of 1977 dollars.
b. Years to aircraft retirement.

planes have long useful lives—as the airlines must do under the noise-charge formula—leads airlines to distinguish retrofits yielding high benefits from those yielding low benefits. Put differently: Under the simple charge, airlines are asked to compare control costs with emissions reductions; but under the formula, costs are compared with noise-exposure benefits. Moreover, the efficiency advantages of the formula would be harder to duplicate with regulations than with the simple noise charge. Under the simple charge, airlines would retrofit most of the second-generation aircraft and pay the charge for the first-generation aircraft, which are much more expensive to modify—a pattern that could easily be added to the regulatory approach by exempting second-generation aircraft from the retrofit rule. The government would have a much harder time determining which aircraft to exempt to match the noise-charge formula; it would try to duplicate the analysis done by each airline, but with much less complete information. Thus, there is a strong argument for a decentralization to the airlines of decisions on which aircraft to retrofit, which is precisely the result under the noise-charge formula.

Evaluation of standards that vary by airport
As the older aircraft become uneconomical to operate, all non–Part 36 aircraft will eventually be removed from the commercial airline fleets. In 1976, the FAA predicted that the number of non–Part 36 aircraft would fall from 1,584 in 1976 to 322 by 1991. Fuel-price increases since 1976 will hasten that turnover. Thus, the issue in extending the Part 36 noise requirements to the existing fleet is not whether aircraft should meet the standards, but when. Under the current regulations, all aircraft must comply by the end of 1984.

One way of adding geographic variation is to allow differences in the percentage of operations to be perfomed by Part 36 aircraft by 1984.[5] The percentage should be set high for airports in densely populated locations (to encourage airlines to introduce Part 36 aircraft quickly) and set low or not at all for airports with few persons under the flight paths. Airports in an intermediate category should receive some incentive to speed up the use of quiet aircraft without the urgency required for the airports in densely populated areas.

Specifying the scheme
A total of 631 airports provide air-carrier service (Civil Aeronautics Board 1976). The Civil Aeronautics Board identifies 156 of these as "hubs"—airports that carry more than 0.05 percent of the total number of passengers. The vast majority of hub airports would receive no benefits from the retrofit program, either because no jets use the airport or because few people live under their flight paths. Table 3.10 compares retrofit benefits, measured by the number of persons who would be removed from the NEF 30 contour, at the airports identified by DOT as receiving the greatest benefits from noise reduction. As these data indicate, the retrofit program would remove 138,000 persons from the NEF 30 contour around Newark Airport, but there would be no benefits around Dulles and Portland. These data permit a crude classification of airports into the noise-reduction-benefit categories given in table 3.11. Ten airports are characterized as "large-benefit": the three major airports serving New York City (J.F.K., LaGuardia, and Newark), the

Table 3.10 Noise-exposure benefits from Part 36 retrofit, by airport.

	Thousands of persons removed from NEF 30	Rank
Atlanta	5.9	17
Buffalo	8.5	15
Cleveland	16.3	10
Denver	19.6	8
Dulles	0	22
J.F.K.	36.7	4
LaGuardia	101.7	3
Logan	32.3	5
Los Angeles	14.3	12
Miami	25.9	6
Midway	10.6	14
Minneapolis–St. Paul	4.4	19
Newark	138.0	1
New Orleans	2.5	21
O'Hare	108.0	2
Philadelphia	17.0	9
Phoenix	7.2	16
Portland	0.0	22
San Diego	15.1	11
San Francisco	20.6	7
Seattle	10.9	13
St. Louis	5.0	18
Washington National	2.6	20
Total	603.1	

Source: Derived from data in U.S. Department of Transportation 1974, pp. 3-7–3-29. This source also presents information on the number of persons removed from the NEF 35 and NEF 40 contours.

Table 3.11 Categorization of airports by retrofit benefits.

Large-benefit (10)
Denver

J.F.K.

LaGuardia

Logan

Los Angeles

Miami

Newark

O'Hare

San Diego

San Francisco

Moderate-benefit (13)
Atlanta

Buffalo

Cleveland

Dulles

Midway

Mineapolis–St. Paul

New Orleans

Philadelphia

Phoenix

Portland

Seattle

St. Louis

Washington National

Small-benefit (133)
All other "hub" airports

Table 3.12 Geographically varying standards: Percentage of Part 36 aircraft operations for three airport categories.

	Airport category		
	Large-benefit	Moderate-benefit	Small-benefit
1976 (actual)[a]	21.3	12.9	23.9
Current 1984 requirement	100.0	100.0	100.0
Alternative 1984 requirement	100.0	75.0	—[b]
Alternative 1990 requirement	100.0	100.0	—[b]

a. Source: Author's calculations, derived from Civil Aeronautics Board 1976.
b. No requirement.

Table 3.13 Percentage of 1976 aircraft operations by airport categories and by aircraft type.

Airport category	Aircraft type			
	High-cost (JT3D)	Low-cost (JT8D)	Complying	All aircraft
Large-benefit	41.6	20.9	23.1	23.4
Moderate-benefit	15.3	17.8	9.5	15.8
Small-benefit	43.1	61.3	67.4	60.8

Source: Author's calculations, derived from Civil Aeronautics Board 1976.

three major California airports (Los Angeles, San Diego, and San Francisco), the major Midwest transfer point (O'Hare), and three major regional airports (Logan, Denver, and Miami).[6] These airports account for 85 percent of the total benefits estimated for all 23 airports. The other 13 airports in table 3.10 are classified as "moderate-benefit." All other airports are put in the "low benefit" category on the presumption that the DOT is correct in its assessment that the 23 airports in its study represent those most benefited by controls on airport noise.

Any timetable to achieve the Part 36 standards in each category is of course arbitrary. Table 3.12 lists a plausible revision of the retrofit rule in which large-benefit airports meet the current timetable, moderate-benefit airports have relaxed deadlines,[7] and low-benefit airports have no requirements.

Costs saved

What would be the cost savings from such a geographically varying scheme? As table 3.13 indicates, the noisier first-generation aircraft are overrepresented at the large-benefit airports, which tend to operate the transcontinental and transoceanic flights for which these planes were designed. Almost 42 percent of the JT3D aircraft operations, but only

Table 3.14 Percentage of aircraft operations by aircraft type.

Aircraft type	Airport category		
	Large-benefit	Moderate-benefit	Small-benefit
High-cost (JT3D)	17.8	9.8	7.1
Low-cost (JT8D)	60.9	77.3	69.0
Complying	21.3	12.9	23.9

Source: Author's calculations, derived from Civil Aeronautics Board 1976.

23 percent of all aircraft operations, occur at the large-benefit airports. To obtain a rough estimate of the number of aircraft to be retrofitted, we can assume that the percentage of retrofitted aircraft is equal to the percentage of operations.[8] Under this assumption, 41.6 percent of the JT3D and 20.9 percent of the JT8D aircraft will be retrofitted in order to meet the Part 36 standards at large-benefit airports.

Determining the aircraft modifications required to meet the compliance timetable for moderate-benefit airports is even more speculative. Table 3.14 shows that only 12.9 percent of aircraft operations at moderate-benefit airports were carried out by complying aircraft in 1976, although the addition of complying aircraft and the retirement of JT3D-powered planes will reduce the percentage of retrofitted aircraft needed. Projecting the number to 1984, table 3.15 indicates that approximately 70 percent of operations will have to be performed by retrofitted aircraft to obtain 75 percent Part 36 operations, if only JT8D-powered planes are modified. Since approximately 18 percent of national operations occur at moderate-benefit airports, an additional 13 percent of these planes will have to be retrofitted by 1983. With the 21 percent compliance requirement for large-benefit airports, approximately 34 percent of the JT8D-powered aircraft would require retrofitting.

The airport-varying standard could thus be achieved by retrofitting 42 percent of the high-cost JT3D-powered aircraft and 34 percent of the low-cost JT8D-powered aircraft. If airlines were able to reschedule their quiet aircraft or eliminate flights to noisy airports at a cost lower than that of retrofitting, however, the actual number of aircraft to be modified would be smaller. At the same time, though, the estimates may understate the need for retrofitting, because they assume that the modified planes operate only at noise-impacted airports when some use at other airports is inevitable. Thus, to err on the side of over-estimating the costs, it is safe to assume that 50 percent of both JT8D-

Table 3.15 Projections of aircraft operations by aircraft type at moderate-benefit airports.

Aircraft	1976 actual[a]	1984 projected
JT3D-powered	75,094	49,877[b]
JT8D-powered	593,600	593,600
In compliance, 1976	99,410	99,410
Forecast in compliance	—	156,693[c]
Total operations	768,104	899,580
Required to be in compliance	—	674,685
In compliance without retrofit	—	256,103
Required operations in retrofited JT8D	—	418,582
Percentage of JT8D operations requiring retrofit	—	70.5

a. Derived from Civil Aeronautics Board 1977.
b. Based on estimates of reductions in four-engine jet aircraft in service, forecast in Federal Aviation Administration 1979, p. 55.
c. Based on national air-carrier forecasts reported in Federal Aviation Administration 1979, p. 66.

powered and JT3D-powered planes would require retrofitting under the geographically varying scheme. Retrofitting half of the noncomplying aircraft would leave at least a 20 percent "surplus" (of operations beyond those required, based on the 1976 figures) for JT3D flights and almost a 50 percent "surplus" for JT8D flights.

Benefits lost

No published estimates of the dollar benefits of retrofitting at various airports are available, but it is possible to calculate the noise-exposure benefits from Nelson's estimate of national retrofit benefits and the airport-specific information provided in U.S. Department of Transportion 1974. Assuming that the average change in exposure level is the same at each airport (which probably understates the benefits at the very noisy airports), residents around the ten largest-benefits airports would receive 85 percent of the dollar benefits for all 23 airports. Weighting decibel reductions according to the illustrative COWPS noise-charge formula (in which retrofitting a JT8D-powered aircraft is one-fifth as beneficial as retrofitting a JT3D aircraft) implies that modifying 70 percent of the JT8D aircraft at the impacted airports will generate approximately half of the total retrofit benefits estimated by the DOT for the other 13 airports. The geographically varying standards would

Table 3.16 Costs and benefits for alternative noise-control strategies (millions of 1982 dollars).

	Costs	Benefits	
		Low	High
4% discount rate			
Geographically varying[a]	606.2	880.7	1,258.1
National[b]	1,212.5	1,074.1	1,534.2
Difference	606.3	193.4	276.1
8% discount rate			
Geographically varying[b]	503.2	639.5	914.2
National[b]	1,006.3	779.8	1,114.9
Difference	503.1	140.3	200.7

a. Calculated as described in the text.
b. Obtained from table 3.6.

thus generate 92 percent of the benefits for the retrofit rule at those 23 airports.

Some benefits will be lost from exempting flights at the low-benefit airports from the retrofit rule. Nelson's (1978) national benefit estimates assume that 12 percent of the benefited population is outside the 23 airports studied by DOT. Although these 12 percent probably receive less than 12 percent of the noise benefits (since the average change in NEF levels due to retrofitting at these smaller airports is undoubtedly less than at the 23 noisy airports), we can conservatively assume that exempting flights at low-benefit airports sacrifices 12 percent of the national benefits. The total benefits of the geographically varying strategy are therefore 82 percent (92/112) of the national benefits.

Costs and benefits compared
The cost and benefit estimates in table 3.16 include results for high and low benefit estimates and for both a 4 percent and an 8 percent real discount rate. The net benefits (benefits minus costs) are shown in table 3.17. The benefits of the geographically varying strategy exceed the costs under all cases, with net benefits ranging from $136.3 million to $651.9 million. In contrast, the added costs of extending the standards to all airports exceed the extra benefits by as much as $412.9 million.

Since most of the assumptions underlying the geographically varying scheme have overstated the costs an understated the benefits, these figures probably provide lower bounds on the difference between the

Table 3.17 Net benefits from aircraft noise controls (millions of 1982 dollars).

	Low		High	
	4%	8%	4%	8%
Geographically varying	274.5	136.3	651.9	411.0
National	−138.4	−226.5	321.7	108.6
Incremental	−412.9	−362.8	−330.2	−302.4

Source: table 3.16.

national and geographically varying strategies. There is little doubt that switching from national standards to a scheme that accounts for differences in noise benefits across airports would add to the national welfare.

Would a charge formula do better?

Although an empirical study of this issue is not possible with the current data, adding geographic flexibility probably strengthens the case for the noise-charge scheme. For one thing, the geographically varying regulations are relatively crude; all airports are assigned to one of three categories, with each group allowed a rather arbitrary percentage of Part 36 aircraft. If noise damages vary substantially within each category, the equal treatment of all airports within each group will generate inefficiencies. In contrast, noise charges can easily vary by airport.

Equalizing the marginal cost of noise abatement is probably even more important in the geographically varying case. As noted above, the advantages of adopting a charge depend upon the extent of variation in the cost per EPNdB of reducing noise emissions. For a national charge, most of the variation is due to differences among airlines in the composition, age mix, and utilization of aircraft fleets. The cost variations are undoubtedly greater under a geographically varying strategy, however, because airlines also have the option to reduce noise by rescheduling aircraft. Some airlines would thus be able to reduce the number of noisy aircraft at noise-impacted airports at very little cost, while others might be faced with retrofitting their entire fleets.

I conclude, therefore, that switching to a noise-charge formula has the potential to increase the net benefits of aircraft noise control considerably. The formula concentrates the costly retrofit requirement on aircraft that generate particularly annoying amounts of noise, fly at night, and use airports in densely populated areas. Airlines have more relevant information on costs and benefits than the government, and the charge scheme exploits this knowledge whereas the standards scheme does not. Although a geographically varying set of standards could generate many of the same improvements, that approach is cruder than the noise-charge formula.

Administrative and Enforcement Costs

A control strategy is worthless if it cannot be administered and enforced effectively. Some regulatory schemes have failed simply because they have become mired in litigation and confusing administration. New approaches such as the noise charge may be particularly vulnerable to administrative and enforcement difficulties.

Proponents and opponents of the charge scheme differ greatly in their predictions of the strategy's administrative ease. Critics argue that the need to monitor individual emissions might result in interminable squabbling over the correct charge; supporters insist that a charge scheme would eliminate the "brinksmanship" and delay that often accompany enforcement of strict standards. There are no examples of charge schemes that can be used as evidence to resolve this question, but a rather strong *a priori* case exists that a noise charge would be less burdensome to administer and enforce than comparable noise standards or marketable-rights schemes.

Advantages of the noise charge
A charge scheme is well-suited to aircraft noise control for several reasons. First, the number of emission sources is relatively small. Compare the task of controlling aircraft noise with that of reducing automotive emissions. The fifteen major domestic and international airlines operate about 2,100 jet aircraft in the United States, with the average number of operations per plane ranging from about 1,500 to 6,500 per year (Federal Aviation Administration 1979). In contrast, there are almost 120 million passenger cars traveling on U.S. roads, and over 143 million licensed drivers making several billion trips per year (Motor Vehicles Manufacturers Association 1980).

Second, existing information is sufficient to determine a reasonable noise charge. The Civil Aeronautics Board now requires airlines to report their annual landings and takeoffs by airport and aircraft type; the FAA could use these data along with the Part 36 noise levels to calculate a noise charge similar to that discussed in the preceding section. Adding the night penalty would require the airlines to submit information cross-classified by time of day, as well. One reason a noise charge can be assessed so easily is that there is general agreement over use of the Part 36 test procedure to measure aircraft noise.[9] In the case of automative emissions, one of the first issues regulators had to address was the test procedure to measure compliance; several versions were developed after the standards went into effect, and the resulting uncertainty contributed to the early confusion over compliance schedules (Jacoby and Steinbruner 1973, chapter 5). In addition, the belief that actual emissions are considerably greater than those of prototype cars somewhat compromises the credibility of control (Mills and White 1978).

Third, even if monitored data were to be used to set the charge, the monitoring could be accomplished reliably and cheaply. A computerized system for processing noise data and matching them with flight information from the air traffic control system is already in operation at Washington National, Dulles, and Logan, and many other airports have monitoring systems that could be upgraded. The capital cost of a complete system is about $200,000, with annual operating costs of about $50,000 (Miller 1979, p. 156).

Fourth, a noise-charge scheme may prevent the conflicts that arise from "all or nothing" controls. A common pattern in environmental regulation has been for the affected industry to complain that standards are too stringent and then for the government to relax or postpone them. This adjustment process has often been haphazard, with regulators threatening shutdowns and industries attacking the validity of any regulation whatsoever (Kneese and Schultze 1976; Mills and White 1978). One danger of this process is that it can generate a midcourse correction that accommodates industry objectives for relief rather than public objectives for reform. As a result, the correction may actually decrease the cost-effectiveness of control.

The recent change in the retrofit rule after a protracted battle in Congress suggests some of the disadvantages of the standards approach to noise control. Following the traditional scenario, the Air Transport Association opposed the original standards, arguing that extending the noise requirements to the existing fleet would not yield benefits commensurate with the costs, the compliance schedule could not be met, and the requirement would compromise fuel efficiency and safety objectives (Federal Aviation Administration 1976). Congressional supporters of tough aircraft-noise standards (also supported by most airport operators) steadfastly refused to fund noise-control measures out of ticket-tax revenues or to relax the standards. After much debate, Congress extended the timetable for retrofitting only for the planes with two and three JT8D engines—those for which retrofitting is most cost-effective. Though this compromise provided relief to many of the smaller airlines, relaxing requirements for the older four-engine aircraft would have been more cost-effective.[10]

Can the administrative costs of standards be reduced?
The enforcement of noise regulations might be simplified if it were to be combined with one of the flexible enforcement strategies discussed above. A noncompliance penalty based on the benefits to an airline of operating a non–Part 36 aircraft would change the "all or nothing" character of regulations. Indeed, it would be relatively easy to devise a noncompliance formula based on aircraft type. The closely related "safety-valve" system, in which a penalty provides a cap on the airlines' cost for exceeding the standard, would also help avoid the industry's resistance to such a large distinction between compliance and noncompliance.

In terms of administration, there are two major distinctions between the noise charge and these modifications of standards. Paying the noise charge carries no implication of wrongdoing, and thus one would expect the airlines to resist the charge less. On the other hand, the airlines would probably pay more under a charge scheme, since they would pay for the decibels they emitted as well as paying the compliance costs for those they eliminated. The prospect of higher payments might make the airlines more likely to attack the legality of the noise charge and engage in nonpayment of charges, disputes over the amount, and the like (see Harrison and Portney 1982).

The burden of administering and enforcing noise standards could also be reduced if the government were to finance retrofitting or replacement of older aircraft. The early plan of the Air Transport Association was to couple support for the Part 36 requirement with a financing arrangement in which the federal government would assume the major burden (U.S. Congress 1976). The proposal included replacing 2 percent of the domestic air passenger ticket tax with a 2 percent tax to subsidize retrofitting and replacement. Though federal financing of retrofitting would reduce the administrative burden of noise control (particularly the likelihood of serious delay or brinksmanship), it would also eliminate any incentive to consider the cost-effectiveness of noise reduction.

Disadvantages of the marketable-rights approach
Administrative burdens may represent the greatest disadvantages of
the marketable-rights strategy. Under the simplest rights scheme, the
FAA would set a total number of EPNdBs. Although in theory setting
the quantity has the same effect as setting the price, several administrative obstacles are likely to arise. (See Hahn and Noll 1982 for a more
complete discussion of these issues in the context of a marketable-
permits scheme for air pollution in the Los Angeles area.) It may take
time to establish market equilibrium, particularly for a commodity such
as airport noise that has been treated as free (or subject to rather specific
regulations) in the past. Studies with much simpler markets have suggested that equilibrium is not reached instantaneously (Plott and Smith
1978); therefore, in the short run at least, the marketable-rights strategy
is likely to generate uncertainty among the airlines. And since noise
rights will be necessary for airline operations, some companies might
engage in strategic behavior, such as purchasing extra rights to escalate
the cost to competitors. Strategic behavior is particularly likely when
competitors in the product market are also competitors in the pollution
market.

Thus, although the noise-rights market may eventually operate with
little government supervision or need for enforcement, the marketable-
rights strategy may be very difficult to administer in the short run. In
contrast, the noise-charge scheme avoids the problems of price uncertainty and strategic behavior.

Flexibility Over Time

Time both simplifies and complicates a regulating agency's task. It
allows midcourse corrections on the basis of better information or a
less frantic political climate, but it also forces the agency to aim at a
moving target. This section considers the relevance of time-related
issues in the choice of an instrument to control aircraft noise. It also
considers whether the choice of a regulatory instrument might affect
the process of change itself, primarily by stimulating the development
of new noise-reduction technology.

Improving the regulations
Advocates of noise charges often argue that the scheme offers a better
mechanism than standards for midcourse corrections. After setting the
charge, the regulator can observe the airlines' reactions, compare the

marginal benefits of slightly more control, and then adjust the charge until the desired standard is met at the least cost. (See Russell 1979 for a discussion and critique of this characterization.) The noise charge is even easier to adjust if the marginal benefits of control are constant or can be specified by a formula that does not change (for example, if the relative annoyance of day and night noise is the same). If benefits are correctly perceived at first, no midcourse correction is necessary; changes in the airlines' view of their costs will be "automatically" translated into noise-control decisions that are efficient. If new information suggests that benefits per EPNdB are different than originally estimated, the government can simply raise the charge without obtaining information on the marginal cost of control.

In practice, these advantages are not likely to be very important. Opportunities for noise reduction involve major modifications that would probably not change if the noise charge were different. If an airline retrofits a 707 when the charge is set at $20 per EPNdB, it cannot reverse the decision if the charge is reduced to $15. In addition, it is not likely that much more information on the benefits of control will be forthcoming; enough studies have been done to suggest that there is a relatively narrow band for average benefits. Finally, since planes manufactured before 1972 are responsible for most noise damage, noise control is largely a medium-range strategy. In the long run, and certainly by the year 2000, almost all aircraft operations in the United States will involve planes that meet or exceed the Part 36 noise standards. In practice, therefore, one would expect the original noise charge to remain in effect for the duration of the effort to quiet the existing fleet.

Changes in costs and benefits
Four major changes may affect the desirable level of aircraft noise: decreases in the cost of aircraft noise control, increases in the number of aircraft operations, increases in the price level that reduce the real noise charge, and increases in the number of persons in noise-impacted areas. Assuming that the noise regulations are likely to remain the same, the question arises of how the various strategies would fare if such changes occurred.

If marginal benefits are constant, the noise-charge scheme will encourage airlines to adjust to the first two changes in a socially optimal way. Figure 3.1a illustrates the response of airlines to a constant noise charge of OA if the cost of controlling aircraft noise decreases or if the number of aircraft operations increases The MC_0 curve represents the initial cost conditions for reducing noise, which would lead cost-minimizing airlines to emit E_A emissions. The MC_1 curve shows a decrease in the cost of reducing emissions, which would occur if, for example, the fuel-economy advantages of the new aircraft were to make replacement a less expensive option. Under the charge scheme, emissions would fall to AS, the optimal level. The marketable-rights scheme has a less favorable effect because price rather than quantity of emissions changes. The price of an EPNdB permit falls to OB. The consumer surplus loss is equal to the shaded area RST.

The marketable-rights scheme has a similar deficiency when airlines want to expand their aircraft operations, either by increasing the use of existing aircraft or adding new aircraft. An increase in desired aircraft operations can be translated into a shift outward in the MC curve, from MC_0 to MC_2, since it is more expensive for the airlines to limit emissions to any particular level. Under the charge scheme, emissions increase to AW, the new optimum when marginal benefits are constant.[11] Restricting the number of emission rights to E_A generates a price rise to OC and a welfare loss equal to the triangle RUW, since limiting emissions now costs the airlines more than the value that households place on the noise reductions.

Setting noise standards for new aircraft allows more flexibility than the marketable-permits scheme, because noise emissions can increase if new planes meet the standards. The result will tend to be less constraining than under the marketable-rights scheme, but probably less flexible than under the noise-charge scheme. In figure 3.1a, a standard that allows aircraft to increase emissions to E_B reduces the welfare loss to VXW.

The noise-charge scheme fares less well when changes occur in the marginal-benefit function, for example when inflation decreases the real value of a noise charge set in money terms. As figure 3.1b illustrates, if over time the real noise charge falls from OA to OB, airlines will increase noise emissions from AR to AS. Since the real value of benefits is assumed to remain the same, the increase in emissions lowers consumer surplus by an amount equal to RST. In this case, the marketable-rights strategy is superior because the number of emission rights remains at AR, the optimal level. It is of course simple to fix the noise charge in real terms, or account for other anticipated changes in the value of noise exposure improvement. For example, the national uniform charge of $7 per EPNdB in 1975 dollars would be $12 per EPNdB in 1982

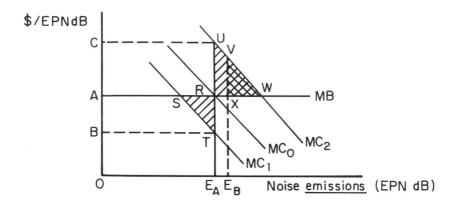

(a) Changes in Costs

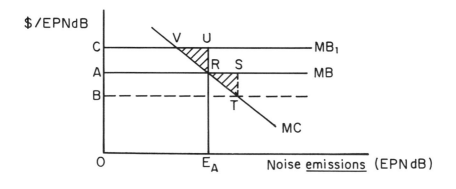

(b) Changes in Benefits

Figure 3.1 Inefficiencies created by changes in costs and benefits under alternative regulatory strategies.

dollars. If, as Nelson (1978) assumes, the real value of noise improvements grows at a rate of 3 percent per year, the value would be $15 in 1982.

The charge strategy and the marketable-rights strategy fail to account for changes in the number of persons exposed to noise levels. The effect of increasing the number of persons living under noise contours is depicted in figure 3.1b as a rise in the marginal benefit curve from MB to MB_1 (that is, the benefits from reducing a given number of EPNdBs are increased). Under the fixed charge and standards strategies, noise reduction remains at CU while the optimal noise reduction is now CV. The inefficiency loss from "too much" noise is equal to UVR in both cases.

Since the efficiency loss consists of noise burdens to new residents, one might argue that such losses should be treated differently from those imposed on residents who moved in before airport noise became a major problem. This argument would entail a different marginal-benefit function, however, because it is likely that the number of newer residents is substantial. Indeed, the largest losers might be those who sold homes with a "noise discount."

The importance of changes in population density over time varies considerably from airport to airport. Some, such as Logan, are surrounded by built-up areas where the population will probably decline. Many others are located in areas where, despite the pleas of airport operators and federal officials, population densities are increasing. (See Blitch 1976 for a discussion of the unsuccessful efforts of the San Francisco Airport to prevent housing development in a noisy area.)

Incentives to Develop New Technology

A common criticism of the standards approach to noise control is that it provides no incentive to develop technology to reduce emissions below the mandated level. In theory, standards do encourage development of less costly technology for meeting a given standard. In practice, however, many emission standards have generated perverse incentives for technological development because the compliance time-table has encouraged polluters to develop relatively costly though more immediate technology. Automobile emissions are often cited as a case in point. Many commentators believe that the 1975–76 deadline established in the 1970 Clean Air Act for a 90 percent reduction in emissions forced the automobile companies to choose high-cost, unreliable, "bolt-on" technology rather than a lower-cost technology that would not necessarily have been ready for production by the deadline (Jacoby and Steinbruner 1973; Mills and White 1978). An emission charge rather than an emission standard, according to this line of argument, would have allowed the auto manufacturers to develop less expensive and more reliable systems because they would not have faced the prohibitive sanctions for noncompliance (a fine of $10,000 per car). In addition, the automobile companies might have developed means of achieving more than a 90 percent reduction in some pollutants.

Mandating a noise charge rather than a noise standard thus promises to speed the introduction of lower-cost noise-reduction technology by providing incentives to develop technology that reduces noise below the standard levels and also providing flexibility for the airlines to develop less costly noise-control techniques even if they cannot be implemented as quickly as a legislative timetable would require. The FAA's cost-benefit analysis of the Part 36 extension suggests that the noise-charge scheme would offer both of these advantages by leading to faster replacement of 707s and DC-8s with new long-distance jets such as the 727-200, the 747, the DC-10, the L-1011, and the A-300. Aircraft manufacturers are also developing new "low-noise" aircraft to meet the stricter Part 36 standards.

Although the FAA did not explicitly consider the effect of a noise charge on the introduction of new planes, it did provide estimates of the noise benefits of replacing rather than modifying the first-generation aircraft. Noise damages over time are shown in figure 3.2 for the baseline case (no additional noise regulation) and three scenarios:

• retrofitting 100 percent of noncomplying aircraft (both JT3Ds and JT8Ds) according to the proposed time schedule,

• retrofitting all noncomplying JT8D aircraft, but only 100 of the JT3D aircraft, and replacing the remainder with new-technology aircraft, and

• retrofitting all noncomplying JT8D aircraft, and replacing all JT3D aircraft with new-technology aircraft.

Retrofitting all aircraft generates larger benefits in the short run, but the options permitting replacement generate much greater benefits after the early 1980s. These large benefits occur because the replacement aircraft were assumed to generate enough demand so that production of the new-technology aircraft could begin in 1981 rather than in 1985. Earlier production has a "snowball" effect on noise benefits because quieter, more fuel-efficient aircraft will be available to meet airline

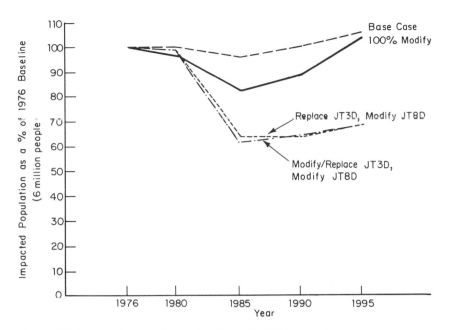

Figure 3.2   Impacted population under alternative FAA scenarios.

demand for additional as well as replacement aircraft. The FAA study thus raises the possibility that a strategy such as the noise charge, which offers airlines the possibility of waiting to reduce noise from the older JT3D engines until replacement is possible, would generate greater long-term benefits than noise standards imposed with a rigid timetable.

Whether airlines pay the charge for JT3D-powered aircraft in the first few years and then replace these planes depends on the size of the noise charge and the characteristics (noise levels, fuel consumption, price, passenger size, route length, and date available) of the new aircraft. My analysis suggests that most airlines would respond to a uniform noise charge of $7 per EPNdB by paying the charge rather than re-trofitting their JT3D-powered aircraft. Under the noise-charge formula, however, retrofitting is more likely because it penalizes the higher EPNdB levels from JT3D aircraft heavily; the formula may therefore result in a slower introduction of new-technology aircraft than the uniform charge. Other regulatory options, such as permitting airlines to count a firm order for the newer aircraft as meeting the standard, may provide greater incentives for more rapid use of the quieter aircraft.

Differential Impact on Airlines

Although most analyses focus on the overall effects of control, the costs and benefits of environmental regulation are generally not distributed equally throughout society. People living near airports will reap the benefits of aircraft noise reduction, while airline users (both passengers and freight shippers) will eventually pay most of the costs of the improvement if no major federal subsidies are available. The airlines will bear some of the costs, although—as demonstrated in this chapter—some may gain when noise levels are controlled.

From a national perspective, the gains and losses for different carriers tend to cancel each other. These differential impacts are nonetheless important in evaluating alternative noise-control strategies, since large variations in costs may be considered unfair. Some airlines have argued that, because the older aircraft met federal certification standards when they were purchased, the Part 36 extension represents an inequitable if not illegal condemnation of airline property. (See testimony in U.S. Congress 1976.) Noise standards for new aircraft do not raise the same fairness objections, because only future decisions are affected. Indeed, it is not clear that noise standards for new aircraft increase the cost of the planes. (National Academy of Sciences 1977 discusses the complementarity of noise and fuel-economy objectives in aircraft design.)

Large variations in costs may also jeopardize the political acceptability of noise control. In principle, firms that cannot cover their full social costs, including the cost of environmental damage, should close down or relocate; in practice, however, these adverse economic impacts are serious obstacles to control. Many of the inefficiencies of current environmental standards can be traced to efforts to mitigate losses (Harrison and Portney 1982).

Finally, the differential impacts of noise-control strategies might influence the competitive nature of the airline industry and thus affect its overall efficiency. If marginal airlines are especially hard-hit or if new entrants to the industry face greater barriers, noise regulations may work to lessen competition. Competitive effects may be particularly important right now as deregulation is being implemented.

Impacts of the Part 36 extension on air carriers

It is not difficult to identify, in general, which air carriers would gain and which would lose from the retrofit rule. The airlines facing the greater costs are those that have large numbers of older, expensive-to-modify aircraft that operate on highly competitive routes dominated by new aircraft. The airlines standing to gain the most are those that operate quiet planes on routes currently served by many older planes; these airlines would obtain windfall profits through fare increases for noise control. A full analysis of how the retrofit requirement would affect airlines' profits would entail disaggregating major routes and predicting unit cost increases (based on decisions to retrofit, replace, or retire aircraft) as well as fare increases. Since airlines now have considerable leeway to enter markets and set fares, the study would also have to predict changes in routes and strategic pricing behavior. (For example, an airline might enter a route dominated by non–Part 36 aircraft to obtain the benefits of the fare increase.) Since such an analysis would be very difficult to complete, this study provides only a crude categorization of likely winners and losers by calculating the costs each airline would incur if all its planes were retrofitted.

The first three columns of table 3.18 list the percentages of high-cost (JT3D-powered), low-cost (JT8D-powered), and complying aircraft in the fleet of each member of the Air Transport Association.[12] Only ten airlines operate the high-cost 707 and DC-8, whereas the low-cost aircraft are much more evenly spread among the companies. Excluding Flying Tigers and Pan American, the fractions of low-cost aircraft in the fleets range from 44 percent for Frontier to 100 percent for three of the smaller regional airlines (Alaska, Aloha, and Hawaiian), with an average of 73 percent. The annual retrofit costs to each airline (column 4) therefore vary a great deal. The smaller airlines generally incur costs of $1 million per year or less, while the major airlines spend more than $20 million. United Airlines is predicted to incur the greatest expense, almost $26 million. Five of the 24 airlines (American, Delta, Pan Am, TWA, and United) account for 72.2 percent of the total costs of retrofitting.

Total cost is obviously a misleading measure of the retrofit program's impact on the airlines, because it ignores fare increases, differences in capacity, and differences in airlines' ability to weather declines in profits. The sixth column of table 3.18 gives retrofit costs per 10,000 passenger miles. Assuming constant load factors, this figure is proportional to retrofit cost per seat-mile, the standard unit for comparing airline costs. Costs per 10,000 passenger-miles range from $1.00 for Continental Airlines to $14.01 for Wein Airways, representing between about 0.2 and 1.0 percent of total operating revenues for most airlines. When

Table 3.18 Differential impact of Part 36 extension on member airlines of Air Transport Association.

	Fleet mix (%)				Cost/10,000 passenger miles	Cost as % of operating revenue	Cost as % of profit
	High-cost[a]	Low-cost[b]	In compliance	Annual cost			
Alaska	—	100	—	$ 315,000	4.69	0.42	9.2
Allegheny	—	84	16	2,800,000	7.69	0.56	17.1
Aloha	—	100	—	378,000	11.44	0.75	24.2
American	33	52	15	23,156,000	9.40	0.97	28.3
Braniff	15	84	1	5,551,000	7.39	0.71	14.4
Continental	—	72	28	1,453,000	1.99	0.22	5.6
Delta	12	76	12	10,453,000	5.47	0.55	9.0
Eastern	—	85	15	7,175,000	3.48	0.35	20.7
Flying Tigers	75	—	25	3,060,000	—	1.07	17.8
Frontier	—	44	56	1,008,000	5.34	0.43	7.3
Hawaiian Airline	—	100	—	315,000	7.29	0.40	49.7
Hughes West	—	93	7	1,400,000	6.88	0.56	14.9
National	—	72	28	1,330,000	2.14	0.25	20.7
North Central	—	53	47	980,000	7.64	0.43	7.1
Northwest Orient	2	60	38	2,737,000	2.47	0.26	3.0
Ozark	—	69	31	1,085,000	8.88	0.56	13.4
Pan Am	45	13	41	10,850,000	6.20	0.55	24.1
Piedmont	—	53	47	745,000	7.49	0.52	12.6
Southern	—	80	20	980,000	9.37	0.61	11.9
Texas Int.	—	87	13	910,000	7.80	0.62	9.8

TWA	48	46	6	22,806,000	9.56	0.99	35.2
United	25	59	16	25,954,000	8.18	0.88	25.4
Western	26	65	9	3,101,000	3.70	0.45	21.6
Wein	—	70	30	294,000	14.01	0.52	(39.6)
Total				$129,018,000			

Sources: Fleet mix, passenger miles, profit derived from Air Transport Association 1978; cost of retrofit per aircraft from Nelson 1978, p. 207.
a. High-cost planes include 720, 707, and DC-8.
b. Low-cost planes include 727, 737, DC-9, and BAC-111.

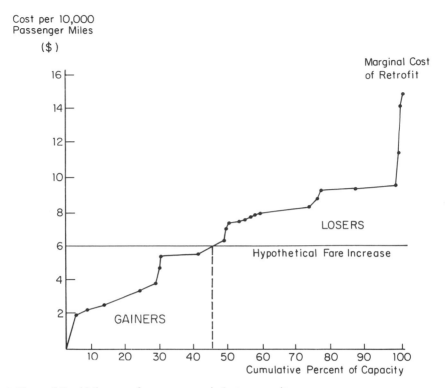

Figure 3.3   Airline retrofit costs versus industry capacity.

expressed as percentage of profit, however, costs vary from 3.0 percent for Northwest Orient to 49.7 percent for Hawaiian Airlines.

To illustrate the net effect of fare and cost increases, figure 3.3 plots cost per passenger-mile against the cumulative capacity of the industry, with airlines arranged in order of increasing cost (Leone and Jackson 1978). Under the retrofit rule, airline fares would increase just as they would if some other factor had risen in price, with the size of the fare increase depending upon demand and competitive conditions. With a hypothetical fare increase of $6 per 10,00 passenger-miles, figure 3.3 shows that the retrofit rule would decrease profits for fifteen of the airlines and increase profits for the other eight. Though the results may differ if separate fare increases are predicted for each travel market, it is clear that low-cost airlines such as Continental, National, and Northwest Orient stand to gain from the Part 36 extension, and that high-cost airlines such as American and TWA are likely to lose.

Table 3.19 Airlines cross-classified by profitability and program cost.

Profit per passenger-mile	Cost per passenger-mile	
	High	Low
High	Piedmont	Northwest Orient
	North Central	Alaska
	Texas Int.	Frontier
	Ozark	Delta
	Southern	Braniff
	Aloha	
Low	Allegheny	Continental
	United	National
	American	Eastern
	TWA	Western
	Wein	Pan Am
		Hughes West
		Hawaiian

Source: table 3.18. "High" and "low" refer to whether the airline is above or below the median.

To identify the airlines that may be most vulnerable to cost increases due to retrofitting, table 3.19 cross-classifies the 23 major airlines by cost and profit per passenger-mile. The "high" and "low" categories refer to whether the airline is above or below the median. The five companies in the high-cost/low-profit category—Allegheny, United, American, TWA, and Wein—are the most threatened by the cost increases. Wein seems particularly vulnerable since it operated at a loss in 1977 and has the highest increase in cost per passenger-mile; of the other four, TWA is the most vulnerable since its profit rate was the lowest and its cost per passenger-mile the highest. Whether retrofit costs would have a long-term impact on the viability of these four large carriers is difficult to determine, but these estimates suggest that these airlines would be the strongest opponents of the Part 36 extension if there is no subsidization of the program. If the government paid for retrofit or replacement, though, these carriers would likely be the strongest supporters of the extension, since they would be able to recover part of the costs of modernizing their fleets.

Differential airline costs under economic-incentive schemes

Because it adds both a cost ceiling for noisy planes and a cost floor for quiet planes, the uniform-noise-charge scheme would likely decrease cost differences among airlines. The $7 per EPNdB represents a ceiling, since an airline can always pay the charge rather than retrofit or retire an aircraft; an airline would, however, have to pay the charge on EPNdBs above 98 for its complying planes. The industry would probably pay more under the charge scheme than under the standards, because the payments for EPNdBs above the threshold would probably overwhelm the savings in retrofit costs.

Introduction of the noise-charge formula might create greater cost disparities among airlines, although not necessarily among close competitors. Cargo airlines, which typically operate noisy aircraft at night between large cities, would probably experience the largest burdens. These differentials, however, might not create changes within particular markets or among competing carriers. For example, cargo airlines operating similar planes on similar routes would all face the same cost increases, and the differential effects would be small. Since passenger carriers operate different planes on very different routes, though, the noise-charge formula would probably generate a greater range of effects among them.

The marketable-rights scheme has greater flexibility than the charge scheme to mitigate cost differentials. Although a charge scheme cannot target income transfers to particular airlines, the EPNdB rights can be distributed to airlines to reduce the disparities. For example, rights distribution can be based on precontrol noise, which would transfer income to those with the greatest demands for rights.[13] Indeed, cost differentials can be made smaller under the marketable-permits scheme than under the regulatory approach. This is another area, therefore, where the charge approach diverges from the marketable-permits system, and one of the very few where the permits scheme has greater advantages.

Revenues from Economic-Incentive Schemes

Economic-incentive schemes can generate significant revenues for the government. These revenues are, of course, transfers rather than net benefits. A noise charge of $7 per EPNdB above 98 would generate about $150 million per year if 50 percent of the existing JT3D-powered aircraft and 100 percent of the JT8D aircraft were retrofitted. A comparable permits scheme would produce about the same amount of revenue if the government were the initial owner of all rights. Although revenues from both control strategies would decrease over time as noisy aircraft were retired from the fleet, a threshold of 98 EPNdB would generate revenue for the foreseeable future, since even the new jets exceed that figure.

The possibility that economic-incentive schemes would become revenue-raising devices probably explains much of the airlines' opposition. But the amount of revenue raised will depend upon the design of the program. Revenue from either the charge approach or the marketable-rights approach could be changed without seriously compromising the schemes' advantages. The government could rebate excess revenue to the airlines according to some formula in which the shares are independent of the noise their planes create (for example, according to the number of passengers). Such a scheme would eliminate excess revenue and still preserve incentives to reduce noise exposure. Revenue from the noise-charge approach could be decreased by raising the baseline above 98 EPNdB or using a combination of subsidies and charges. Under the marketable-rights scheme, revenue could be eliminated simply by allocating the rights rather than auctioning them off.

From a national perspective, collecting revenue might increase the overall efficiency of the economy. Putting revenue into the general treasury is equivalent to reducing other taxes. Since the noise-charge and permits schemes tend to decrease rather than increase distortions in the economy (in contrast to virtually every other tax), adopting such programs would be an improvement. For example, reducing income taxes would reduce current disincentives to work. Without knowing which taxes the revenues might reduce, it is difficult to determine the quantitative importance of this benefit; it may, however, be substantial (Terkla 1980).

Charge or permit revenue may be used for noise-related purposes rather than for tax reduction. The most common suggestions are to earmark revenues for noise monitoring, soundproofing, purchase of severely noise-impacted property, and compensation (Council on Wage and Price Stability 1977). Earmarking is generally a bad strategy, however, because it may disguise costs and lead to the funding of inefficient activities. As discussed in chapter 2, strategies to compensate landowners for aircraft noise either by soundproofing, purchase, or direct payment may be inefficient and even inequitable. Whether soundproofing or property purchases are justified in an efficiency sense will probably depend on the particular circumstances. Several studies have concluded that soundproofing of residences is not one of the best noise-reduction techniques because it does not eliminate outdoor noise and is expensive

(U.S. Department of Housing and Urban Development 1972, pp. 221–228); it may be more cost-effective now because it also reduces heating requirements. The wisdom of purchasing land around airports will depend in part on whether the site can be reused. Whether payments to airport neighbors compensates for losses society should bear depends upon whether true losers can be targeted and what sharing of costs between airport neighbors and airport users is considered fair.

Creating new programs with noise-charge revenues would increase the administrative costs of the scheme, particularly if they were developed and run by federal officials. Administrative costs would probably be lower if, as is likely, individual airports administered the programs. Separation of revenue-raising from revenue-spending authority, however, might increase the likelihood of waste in allocating "found money."

In summary: Revenue from noise-charge or marketable-permit schemes would be a mixed blessing. Though the new revenue might permit the reduction of some distorting taxes, its uses might also create even greater waste. At the very least, airlines would vehemently oppose the schemes if large sums were involved.

Chapter 4

Noise-Control Strategies for Individual Airports

In its policy statement on noise, the Federal Aviation Administration
(1976, p. 2) points out that "the primary obligation to address the
airport noise problem has been and remains a local responsibility." As
discussed in chapter 2, however, federal constitutional and legislative
mandates constrain the control options airport operators and local governments can adopt. They are also constrained by powerful conflicting
interests: Some people want the airport to be a thriving gateway to
commerce; others want it to be a good neighbor. The noise-charge
scheme may lessen the contention. Charging airlines for the damage
each of their flights causes should decentralize—and thereby defuse—
decision making. These airline adjustments are also likely to provide
the cheapest means of achieving noise reductions. Though the federal
government may be loath to set complicated charges for every facility,
an individual airport can build a charge scheme on its existing landing-
fee system. This chapter discusses the advantages and disadvantages
of a single airport's introducing a noise-charge scheme. To illustrate
the milieu in which noise control decisions are made, the section begins
by describing the evolution of noise-control regulations at Boston's
Logan Airport.

Logan Airport's Actions to Control Noise

The Massachusetts Port Authority (Massport) operates Logan Airport
in Boston. Although the airport is located on the coast and some of
the runways point toward the ocean, Logan is one of the most noise-
impacted airports in the country because it is adjacent to a densely
populated area. The Department of Transportation (1974) estimated
that aircraft noise affects about 431,000 persons in Boston.

Logan's current noise-abatement rules came into effect on January
1, 1977. The major debate began in 1973, when Massport adopted a
"fleet noise rule" stipulating that 50 percent of all aircraft using Logan
meet the Part 36 standards by December 31, 1976, and that all aircraft

meet the standard by December 31, 1979. The proposal included a $500 fine for each operation below these percentages. The Logan Airport Master Plan Study, completed in September 1975, lauded the fleet noise rule and recommended that Massport "enforce this regulation to the full extent of its legal authority" (Massport 1975, p. II-45). Massport adopted a Master Plan for Logan Airport in April 1976 (with the Master Plan Study as a technical supplement), which set policy guidelines for the growth and use of Logan—including several commitments to reducing noise impacts. The commitments included establishing a "higher cost to aircraft not meeting Part 36 noise standards" and considering "introducing a noise component into landing fees." The stage was thus set for the development of economic incentives to reduce noise.

The most controversial noise proposal in the Master Plan was for a curfew. Soon after publication of the Master Plan, an evaluation of the curfew concluded that it would reduce the number of persons in the NEF 30 contour by 25 percent (from 76,000 to 57,000). The employment loss, including direct and indirect effects, was estimated to range from 3,600 to 13,000 jobs. Potential sales losses to New England businesses were put at between $81 million and $373 million per year. Both sides of the debate disputed these results. In August 1976, the executive director of Massport recommended that the curfew not be imposed:

In addition to these immediate costs, there are lost opportunities. It seems clear, for example, that where airlines would keep maintenance operations at Logan, the effect of a curfew would be to reduce future growth in maintenance jobs. . . .
There are also long-term effects. While many major users at Logan indicated they could make adjustments if service patterns were changed, they did state that the long-term effects of such changes would influence business decisions on further expansion and location of new facilities with a commensurate impact on the number of jobs created or eliminated. . . .
Beyond all this, there is the amorphous, but nevertheless real, 'business climate' argument, and this is important to the people that make decisions as to plant location or relocation or expansion. Also, in an economy with a 7.8 percent unemployment rate, the prospects for reemployment of people who might lose their jobs if a curfew was imposed are dim at best. The people who might lose their jobs because of a curfew are not those in the high-income brackets but rather they are heads of households who support two, five or even more people in their families on salaries ranging from $5,900 to $18,000.

It is concern for the general economic health of this region that impels me to make this recommendation. There is no question in anyone's mind that Logan Airport is an essential element in the economic vitality of Boston and the New England region. Its ability to serve the area must be maintained. (Massachusetts Port Authority 1976)

The executive director recommended instead a set of alternative noise-abatement measures, which led to promulgation of the regulations passed by the Massport Board in November 1976, to go into effect on January 1, 1977.

The Logan Airport Noise Abatement Rules and Regulations consist of nine articles.

Article I: Aircraft noise proposals
This article prohibits new aircraft types not currently operating at Logan that do not meet the Part 36 noise standards (for example, the Concorde) from using the airport except under emergency conditions.

Article II: Noise-abatement scheduling rule
Replacing the fleet noise rule described earlier, this regulation requires each airline to maintain the "maximum feasible" percentage of Part 36 operations each month rather than a specific compliance schedule. The airlines must submit to Massport's executive director plans that demonstrate that their proposed monthly compliance ratios are the maximum feasible; the executive director then reviews the plans and may prescribe higher ratios if he believes them to be feasible. If an airline owns no Part 36 aircraft, it can operate at Logan without paying a penalty. Failure to meet the monthly compliance ratios is a violation of the regulation and subjects an airline to penalties outlined in article IX. In 1978, one airline had two violations and paid fines totaling $2,850.

Article II also contains target compliance ratios (84 percent by 1979, 100 percent by 1980) that have no legal significance but are widely misinterpreted as a standard that the airlines must meet. (See, for example, Nelson 1978 and Jaynes 1978.) The regulation appears to have been successful in attracting Part 36 aircraft to Logan; from 1976 to 1979, Part 36 aircraft rose from 24 to 32 percent, a 33 percent increase. Over the same period, the percentage of Part 36 operations increased from 26 to 46 percent in 1979, or by 77 percent. Individual compliance ratios ranged from 0 to 100 percent.

Article III: Late-night aircraft restrictions

This article prohibits operation of non–Part 36 aircraft during late-night hours. Under the regulation, the late-night period was expanded from 1:00–6:00 A.M. in 1977 to 10:00 P.M.–7:00 A.M. in 1981. There are four major exemptions: aircraft using Logan for short-term maintenance or inspection that cannot avoid flying at night, operations delayed for a "satisfactory" reason, "necessary" and "unanticipated" equipment substitution, and cargo services that can demonstrate that every reasonable effort was made to use Part 36 aircraft, that the service is "essential to the economy of New England or a part thereof," and that the flight could not reasonably be scheduled outside of the late-night period. The exemption for cargo operations is the most important, since such flights accounted for about 40 percent of operations between 10:00 P.M. and 6:00 A.M. in 1974. During the first seven months of 1978, 56–68 percent of flights between 12:00 P.M. and 6:00 A.M.occurred in Part 36 aircraft. Fines totaling $8,300 were assessed in 1978 for 28 violations of this article.

Article IV: Noise-abatement ground procedures

This article imposes nighttime restrictions on engine runups, training flights, and the use of ground power and auxiliary power units. It also requires the use of certain procedures relating to taxiways and runways at all times. Though the regulations are quite specific, they do allow the airport manager considerable flexibility in granting exceptions. For example, engine runups are not prohibited if the airport manager approves the times and places in advance.

Article V: Certain ground movement by jet and turboprop aircraft not to be conducted by self-propulsion

With some exceptions, this article prohibits self-propelled ground movements and thus requires that aircraft be towed to and from runways and hangars. The compliance schedule called for this rule to apply to all movements in 1979, but the requirement to tow to and from runways is not yet in effect pending the outcome of an FAA study on safety implications.

Article VI: Operating restrictions

This article establishes a nighttime preferential runway system designed to encourage takeoffs and landings over water whenever certain safe conditions prevail. Superseding the original procedure in September

1977, a demonstration program still encourages operations over water but permits pilots to choose another runway if they consider the preferential runway unsafe.

Articles VII-IX
These articles require airlines to submit certain flight data to Massport, stipulate that the various regulations are legally separable (that other provisions are still valid if one is judged illegal), and set penalties for violations of each article. In 1978 there were 35 violations of the noise rules; airlines paid fines of $11,440 plus a 25 percent surcharge.

In addition to promulgating the noise-abatement rules and regulations, Massport has worked with the FAA to develop and evaluate alternative flight tracks to reduce the noise impact in some communities. But the FAA has ultimate authority to prescribe flight tracks. After considerable debate, the FAA recently chose a relatively noisy flight path—a decision Massport criticized as insensitive to airport neighbors.

Massport has not carried out its commitment to charge higher fees to aircraft not meeting Part 36 noise standards. Although officials in the noise-abatement office have considered introducing a noise component into landing fees, no formal action has been proposed. The noise regulations promulgated by Massport thus contain no economic incentives, despite the promising language in the 1976 Master Plan.

Advantages of the Airport Noise Charge

The analysis of the national noise charge in chapter 3 identified the likely advantages and disadvantages of the scheme relative to regulations. Four major advantages stand out.

More cost-effective control
The efficiency advantages of the national noise charge arise from the variability in marginal control cost—among airlines and among noise-control options—and the difficulty that the regulatory authority has in identifying and requiring the low-cost options. By these criteria, an individual airport's noise charge should be even more cost-effective than the national schemes, because airlines have a greater range of control options at a single airport and because an airport operator would find it even more difficult to determine which combination of measures would minimize control costs.

The case of Logan Airport illustrates the promise of a local noise charge. Table 4.1 lists current landing fees for various aircraft and how they would change under the noise-charge schedule developed by the Council on Wage and Price Stability. The charges should provide strong incentives for airlines to substitute quieter operations for noisier ones, reschedule night flights to the daytime, and eliminate the route segment altogether.

The noise charges are likely to induce the use of more L-1011s and DC-10s, and fewer 707s and DC-8s, at Logan. The large penalty on night operations might result in a virtual curfew. It is impossible to estimate the response of airlines to any particular charge, however, because the adjustments depend upon the characteristics of the airline's fleet and the profitability of the flights and flight segments. The noise charge would be one more component in the complex process of scheduling and allocating aircraft. The Massport regulations, though, may not result in efficient scheduling and allocation decisions. For example, the prohibition on non–Part 36 aircraft for night flights allows four major exceptions, which may or may not exempt flights where noise-reduction benefits are small and compliance costs are large. The requirement that airlines maintain the "maximum feasible" percentage of Part 36 operations each month is even less likely to guarantee cost-effective noise abatement. Since aircraft allocation can be a complicated process, it might be impossible for airport operators to determine the costs to each airline of increasing the use of quieter aircraft.

The airlines might also modify landing and takeoff procedures to reduce noise charges. The noise charges in table 4.1 are based on noise measured in the Part 36 compliance tests; with monitoring data, however, an airline could get credit for operating its aircraft in a less noisy fashion. For example, an airline could instruct its pilots to reduce takeoff thrust or to make two-segment landings. If the noise charge varied by runway (as it should), the airline could also request a less noise-impacted runway. The contribution of these changes to cost-effective noise control might nonetheless be small; because these procedures involve safety risks, they are heavily regulated by the FAA and carefully criticized by the Airline Pilots Association. Indeed, many local noise regulations— including those at Massport—defer to pilot discretion when safety is an issue.

A single-airport noise charge might also encourage airlines to take initiatives in reducing the sensitivity of neighbors to noise. Airlines might pay for some soundproofing or barrier construction, for example,

Table 4.1 Influence of noise charges on landing fees at Logan Airport.

Aircraft type	Maximum landing weight (lb)	Approach noise[a] (EPNdB)	Takeoff noise[a] (EPNdB)	Current landing fee[b] ($)	Noise charge for takeoff and landing[c] ($)			
					Day-day	Day-night	Night-day	Night-night
707-320B	247,000	117	113	219	728	5,496	3,422	8,190
727-100	137,500	110	100	122	205	2,012	657	2,464
727-200	150,000	108	101	133	172	1,541	691	2,060
737-200	103,000	109	92	91	143	1,716	143	1,716
DC-8-50	217,000	117	115	193	763	5,531	4,377	9,145
DC-9-30	110,000	107	96	98	108	1,300	108	1,300
DC-10-10	363,000	106	99	323	130	1,168	525	1,561
L-1011	368,000	103	98	327	93	778	436	1,121
747-200	564,000	106	107	501	202	1,240	1,394	2,432

Source: Miller 1979, p. 109.
a. Approximate; based on Federal Aviation Administration 1978.
b. Calculated using landing-fee rate at Logan for fiscal 1977.
c. Calculated using charge schedule given in table 3.2.

if the charge were to be reduced accordingly. The need for coordination among airlines may still limit these actions in practice (as argued above for national charges), but the airport may be able to overcome some of the difficulties by suggesting particular options and clarifying the amount of the charge reductions.

Finally, airlines might simply pass the noise charge on to passengers through higher fares, which would in turn reduce noise by lowering the number of air travelers and hence the number of aircraft operations. Although changes in flights are difficult to predict from aggregate fare elasticities, high surcharges on flights into a given airport are likely to induce substantial changes in traffic. If New York is a close substitute for many transatlantic travelers, for example, a noise charge at Logan might result in substantial diversion of traffic. Similarly, large increases in fares for night flights should cause both freight shippers and passengers to choose daytime schedules.

Noise benefits

A charge scheme such as the COWPS plan for Logan would almost surely induce considerably greater reductions in noise exposure than any regulations promulgated or proposed by Massport. One reason is simply a corollary of the fact that such a scheme is more cost-effective than regulatory alternatives: Since airlines would be spared the very highest cost adjustments, such as eliminating all night flights under a curfew or operating only Part 36 aircraft at all times, an airport should be able to achieve greater reductions in noise exposure for the same cost to the airlines. But the damage-based charge is also inherently more stringent than regulations that major airports in the United States have actually adopted.

Since the airlines' reactions to the charge cannot be predicted with certainty, estimating the precise noise benefits (either in dollar values or in reductions in the number of persons exposed to higher NEF levels) of the scheme is also difficult. Decreases in noise exposure around the airport, however, should substantially reduce the pressure for noise abatement or airport noise liability—the number of persons subject to high noise exposure levels would obviously be lower, and the airport could argue that the noise that remains under the noise-charge scheme represents a reasonable compromise between the airport's rights to operate and the neighbors' rights to be free from noise.

Revenue potential

Noise charges promise additional revenue to the airport or the municipality that operates it. Additional revenue would no doubt be very welcome, especially to local governments that are under strong pressure to reduce property taxes but to maintain services. Noise-charge revenue would be particularly attractive because many of those paying higher air fares or receiving fewer airline dividends as a result of the noise charges would live outside the city. Finally, substituting the noise-charge revenue for other revenue sources would eliminate some of the distortion created by the local government's revenue collection.

The precise amount of revenue collected under the noise-charge scheme would depend upon the charge schedule and the airlines' reactions. The amount may not be trivial. Applying the COWPS noise-charge schedule to 1977 activity at Logan would result in total revenues of approximately $38 million. Logan's 1977 revenues totaled $48 million (Massachusetts Port Authority 1978, p. 65), so the noise-charge receipts would have increased the facility's total revenue by almost 80 percent.

It is not clear that airport operators would welcome the increased revenue. Discussing this potential difficulty in the context of a marginal-delay cost pricing scheme for the New York airports, Carlin and Park (1970) suggest that "a measure is financially acceptable if it meets the financial requirements of the airport operators without an almost equally unacceptable 'embarrassment of riches.' " They estimate that a full marginal-cost pricing scheme would yield approximtely $150 million for Kennedy Airport alone. The "embarrassment" this large increase causes is due to the airport operators' traditional view of their role as a supporter of regional development rather than as a revenue generator.

Raising revenue through a noise charge would be a significant departure from current practices of pricing and financing airports. The federal government has long helped to fund the construction of airfields and related facilities. The most common method of financing the non-federal costs is for the city to sell revenue bonds underwritten by the airlines, a scheme developed by the city of Chicago and the airlines in the 1950s to finance the construction of O'Hare International Airport.[1] The airlines pledge that if airport income (primarily from concessions for parking and rental cars and from terminal rentals) falls short of the total necessary to pay off the principal and interest on the bonds, they will make up the difference through landing fees based on weight. Landing-fee revenue is therefore not directly related to the operating cost of the airfield, and it may represent only a small portion of total revenue.

Both the airport operator and the airlines would perceive a change in the landing fees to a major revenue-raising device as a major departure from their original purpose.

Like the national charge, the revenue-raising potential of a local noise charge may lead to pressure to use the funds for noise-related purposes. The possibility of projects that could reduce noise damage does not mean, however, that the specific options are worthwhile or that the airport operator wants to undertake them. Besides being a relatively inefficient way to reduce noise damage, seeking projects such as sound-proofing or land acquisition on which to spend money would also be a departure from the current cost-recovery philosophy under which airports operate. Such activities would likely be viewed as diversions from the major purpose of the airport, which is to provide safe, reliable intercity transportation.

City or state governments might relieve airport operators of any revenue embarrassment. Since most airports are established by local governments, the city could presumably appropriate revenue in excess of airport operating costs. The state might also obtain some of the revenue by levying an excise tax on the airport's revenues. For example, in 1979 Massachusetts levied a 5-percent excise tax on Massport's gross revenues, which increased Massport's expenses by about $3.5 million per year. Although city or state appropriation of revenue would relieve airports from having to administer soundproofing and other programs, airport operators are likely to oppose (as Massport did) diversion of airport revenues to local or state treasuries. The airlines and their customers would likely join the airport operators in arguing that generating revenues in excess of costs not only is unfair but decreases the vitality of the region. Even local or state governments may view the revenues as an inconvenience if they entail compensating noise-impacted airport neighbors or reducing remaining noise damages.

The preceding suggests that it is difficult to determine whether revenue potential can be viewed as an important benefit of the noise-charge scheme. On balance, local and state governments would probably welcome an additional revenue source, particularly governments in growing areas where additional charges would not seriously affect the region's competitive position. In contrast, airport operators are likely to regard extra revenue from a noise-charge scheme as a disadvantage, or at best a mixed blessing.

Encouragement of competition

Regulations on aircraft noise run the risk of blunting the recent federal mandate to increase competition in the airline industry. A crucial element of deregulation is that new companies have the ability to enter any airline routes. Over time, this should lower fares and bring more responsive service. Removal of Civil Aeronautics Board control over routes and fares has already increased traffic at many major airports (Meyer et al. 1981), which in turn has generated greater annoyance to airport neighbors. Local pressure to contain congestion and noise has thus increased.

As discussed in chapter 2, a California state law now requires that airport operators take steps to reduce and eventually eliminate serious noise impacts. The Civil Aeronautics Board has objected to noise-control measures proposed at three California airports—a San Diego moratorium on use of the airport by new carriers, a San Francisco requirement that all operations by new carriers use Part 36 aircraft, and a Burbank restriction on additional air-carrier operations—because such proposals would restrict competition. Unlike these regulations, a noise charge would apply equally to all carriers, new or incumbent. The airport would be accessible to anyone willing to pay the charge. Airlines with noisier planes would have to pay more for the right.

The marketable-rights approach is also consistent with the spirit of airline deregulation. The CAB and the FAA are exploring the use of a noise-allocation scheme very similar to the marketable-rights scheme discussed in this chapter, and have suggested it as an alternative to the restrictive regulations the California airports proposed. An incumbent carrier wanting to add flights above its allocation would have to reduce the noise of its current flights or purchase rights held by another carrier. A new entrant would have to purchase noise rights from existing carriers. (There is the danger that incumbent carriers would act to restrict competition, especially if the rights strategy were introduced at a single airport.)

Disadvantages of the Charge Scheme

The analysis of the national charge in chapter 3 suggests that the strategy has two potential disadvantages related to administrative feasibility and differential impacts on airlines. In the local context, two other problems arise: Airlines' freedom to choose the cheapest means of controlling noise may decrease traffic and therefore jeopardize the airport's role in maintaining a healthy local economy, and current constitutional and contractual arrangements between the airport and the airlines may create legal obstacles to instituting a noise-charge scheme.

Administrative costs

All charge schemes face the possibility of being overwhelmed by administrative difficulties. Can emissions be monitored? Can the charge be set at an appropriate level? Will firms react to the charge by simply contesting it instead of determining the least-cost combination of controls and emissions? Chapter 3 suggests that such administrative obstacles were not very great in the case of aircraft noise control. Available data provide a good estimate of the marginal benefits of reducing aircraft noise, and the noise case does not involve placing a value on human life and limb. Moreover, using prototype noise levels for various aircraft obviates the need for monitoring.

A charge that varies by noise level, time of day, and runway could be implemented and enforced by requiring that the airlines report their flight operations, with spot checks to make sure the submissions are accurate. The noise charge due could simply be calculated and the airlines billed. Basing the noise charge on monitored data rather than prototype data would be preferable, partly because the prototype data may not accurately reflect actual noise levels in the particular communities of concern. Simpson (1979) and Jaynes (1978) have shown that monitoring can produce more reliable noise data than Part 36 prototype measures.

Designing a more complicated and more efficient charge scheme raises a set of additional questions: Should the charge schedule vary from runway to runway? Should the charge be assessed according to the noise level recorded at a single microphone, or an average level from two or more microphones? What noise level at each microphone corresponds to an equal contribution to noise in the community? Airport operators without sophisticated planning staffs might find these questions formidable, but technical assistance from the federal government might make operators much more willing to establish a noise-charge scheme. Once the scheme was established, administrative requirements to maintain it would be modest.

Uncertainty

The noise-charge strategy entails some uncertainty over the level of control eventually achieved and the means the airlines will use to reduce noise. In terms of national economic efficiency, these uncertainties are not disadvantages because they will be resolved optimally once the airlines respond to the noise charge.[2] To the airport operator, however, both of these uncertainties represent large disadvantages in efforts to balance competing local demands for noise abatement and a high quality of air service. The airport operator is not indifferent to the means airlines employ to respond to the noise charge. Faced with two options to achieve a certain reduction in noise impact, one involving quieter planes and the other fewer flights, the airport operator will choose the quieter planes regardless of how expensive this alternative is from the airlines' or the nation's perspective. Airport operators are therefore always among the strongest supporters of federal noise standards or noise-reduction requirements that generate noise benefits without threatening to reduce air travel substantially. (See testimony of Massport officials in Congressional hearings on noise abatement, cited in Miller 1979, p. 70.)

The example of Massport's noise restrictions tends to support this characterization of the airport operator's perspective. Massport has typically taken noise-abatement actions that have had fairly predictable effects on noise reduction and air service. For example, Massport's requirement that airlines use the "maximum feasible" number of Part 36 aircraft based on the airlines' records provides a way of achieving noise reductions without sacrificing the number of operations. Even the airports in California that have proposed to prohibit future flights are probably attracted to the certainty of a ban, which is nonetheless an inefficient means of maintaining current noise levels. One of the reasons a marketable-rights scheme has more appeal to airport operators than a noise charge is that setting the quantity provides greater certainty, even if the precise levels of noise or aircraft operations are set arbitrarily.

The disadvantages of a noise charge from the airport operator's perspective are the most serious if he acts alone, because the impact on service will be greater than if all competing airports acted together. Indeed, part of the reason a noise charge is likely to generate cost-effective noise control is that airlines could redirect some of their flights to other airports. Most airports would not, however, voluntarily impose noise charges that put them at a competitive disadvantage. Though an economist might correctly point out that any service lost is of less value than the noise damage it causes, or that changes in economic activity are likely to be transfers from the standpoint of society as a whole,[3] a brief look at the literature on local noise control and air fares indicates that airport operators and local businesspeople perceive positive net spillovers associated with airport activity. Even those who suffer from noise may question the wisdom of reducing the annoyance if it threatens economic activity. Local supporters of a curfew at Logan Airport agreed that the plan should be rejected if it would generate substantial employment losses (perhaps because many of them had jobs at Logan). For those whose jobs are not tied to the airport, stringent noise control is more attractive. Few of the neighbors of the Los Angeles Airport whose intense legal efforts resulted in the airport's purchase of large tracts of surrounding land were airport employees. Whether the net effect of airport-associated externalities is positive is irrelevant to the discussion here; it is sufficient that nearly everyone, except perhaps the people suffering from the noise, perceives positive economic spillovers. A noise-control plan that does not appear to be sensitive to this issue will surely be looked upon with disfavor by the airport's operators and by many people in the community who have influence over them.

Increased burden on air carriers
Airport operators see the current weight-based landing fee as a generally equitable means of charging airlines for their use of an airfield. A noise charge would change the distribution of costs among airlines, a result that the airport operator is likely to consider a disadvantage. For example, in describing why they arrived at a compromise solution to the peak-hour pricing problem, Little and McLeod (1972) of the British Airports Authority stated

Any move to a pure passenger/mileage based charge would have greatly disturbed the relative existing charges paid by different airlines, and also by charter flights as compared with scheduled traffic. There is a lot to be said in favour of steering one's customers gently.

To illustrate the impact that a noise-charge scheme might have on fees, table 4.2 gives the noise charges that would have been levied at Logan Airport in fiscal 1977 if the COWPS schedule had been applied. Under the noise charge, individual airlines would have paid 54–1,131 percent higher fees. (The figures overstate actual fees, since the airlines would have adjusted schedules and aircraft in response to the charges.)

Assuming the addition of noise charges, table 4.3 shows how each airline's share of total fees paid would change. American, United, and Flying Tigers—the airlines with the highest proportion of DC-8 and 707 operations at night—would experience the greatest increases. American's share of total fees would grow from 16.4 to 23.9 percent,

Table 4.2 Noise charges at Logan Airport for fiscal 1977.

	Landing fees	Noise charges	Total	Percent change
American	$ 2,173,996	$ 9,984,829	$12,158,825	+459
Allegheny	1,209,183	4,167,095	5,376,278	345
Delta	2,620,628	5,463,467	8,084,095	208
Eastern	2,357,921	6,705,734	9,063,655	284
Flying Tigers	132,298	1,496,284	1,628,582	1131
National	95,372	128,676	224,048	135
North Central	56,988	69,552	126,540	122
Northwest	142,295	231,253	373,548	163
Pan Am	353,202	242,414	595,616	67
Seaboard	68,367	158,688	227,053	232
TWA	1,604,264	4,318,002	5,922,266	269
United	945,731	3,635,090	4,580,821	384
Air Canada	184,256	196,560	380,816	107
Air France	83,738	84,032	167,770	100
Alitalia	75,322	40,560	115,882	54
British Airways	262,781	147,056	409,837	56
Aer Lingus	72,904	113,568	186,472	156
Lufthansa	247,311	178,464	425,775	72
Swissair	249,745	355,082	604,827	142
TAP	46,789	37,856	84,645	81
Air New England	162,625	—	162,625	—
Commuter carriers	112,028	—	112,028	—
Total	$13,257,744	$37,754,260	$51,012,004	285

Source: Miller 1979, p. 139.

Table 4.3 Airlines' shares of fees at Logan Airport for fiscal 1977.

	Percent of landing fees	Percent of landing fees and noise charges	Change (increased percent burden)
American	16.4	23.9	+7.5
Allegheny	9.1	10.5	+1.4
Delta	19.8	15.9	−3.9
Eastern	17.8	17.8	−
Flying Tigers	1.0	3.2	+2.2
National	0.7	0.4	−0.3
North Central	0.4	0.2	−0.2
Northwest	1.1	0.7	−0.4
Pan Am	2.7	1.2	−1.5
Seaboard	0.5	0.4	−0.1
TWA	12.1	11.6	−0.5
United	7.1	9.0	+1.9
Air Canada	1.4	0.8	−0.6
Air France	0.6	0.3	−0.3
Alitalia	0.6	0.2	−0.4
British Airways	2.0	0.8	−1.2
Aer Lingus	0.5	0.4	−0.1
Lufthansa	1.9	0.8	−1.1
Swissair	1.9	1.2	−0.7
TAP	0.3	0.2	−0.1
Air New England	1.2	0.3	−0.9
Commuter airlines	0.9	0.2	−0.7

Source: Miller 1979, p. 140.

United's from 7.1 to 9.0 percent, and Flying Tigers' from 1.0 to 3.2 percent.

The revenue that the charge generates could be kept constant by providing bonuses for flights below some EPNdB level and charging for flights above that level. Such a subsidy plan, of course, creates problems of deciding about bonus payments; if an airline cancels a flight, should it receive the noise bonus? The same differential impact on the various airlines, however, would still occur; there is simply no way to keep noise charges from having their most serious effect on smaller airlines that operate relatively noisy aircraft at night. These airlines generally have few if any quiet aircraft, and thus they would be unable to reduce their liability by reallocating their planes. Although one can argue that these differential impacts are fair because they reflect the differential noise damages imposed by various airlines, airport operators are likely to resist such large changes in the shares of total fees paid. In discussing noise-based landing fees, the Logan Master Plan Study (Massport 1975) addressed the impacts on smaller carriers:

There is an obvious problem of finding ways to retain the marginal carriers, without violating the requirement that the charge be nondiscriminatory. One possible solution would be to increase the landing charges across the board but simultaneously to offer inducements to those carriers in need of them, pehaps in the form of reduced terminal rentals.

Such a subsidy plan, however, tends to defeat the purpose of the noise-charge scheme.

The variation in costs among airlines under an economic-incentive scheme may be lower than under some other equally stringent regulation such as a curfew or a noise maximum. Chapter 3 pointed out that a national noise-charge scheme would reduce the differences in control costs relative to national standards. It is clear, moreover, that an appropriately designed marketable-rights strategy could avoid large changes in the shares that airlines pay. By providing the marketable rights disproportionately to airlines with noisy operations, an equivalent noise reduction could be achieved without changing the distribution of welfare among current airport users. Indeed, this feature of the marketable-rights approach may explain much of its popularity among airports thatabl now considering economic-incentive schemes.

Legal obstacles

There is little doubt that the federal government has the legal authority to impose noise charges, but noise charges imposed by an individual airport would almost certainly face legal challenges from the airlines as a violation of the constitution, of state and federal laws, of international treaties, and of contracts between the airport and the airlines. (See Baxter and Altree 1972, Rosenthal in Massport Master Plan Study Team 1975, Nierenberg 1978, and Miller 1979 for a discussion of these laws.) Current contracts between airports and the airlines appear to present the biggest legal hurdles.

In *Evansville-Vanderburgh Airport Authority District* v. *Delta Airlines, Inc.*, the Supreme Court upheld certain local user fees after applying a three-part constitutional test:

• Does the fee "discriminate against interstate commerce"?

• Is the fee assessed according to "a fair, if imperfect, approximation of the user of the facilities"?

• Is the fee "excessive in relation to costs incurred by the taxing authorities"?

Nierenberg (1978) concludes that a noise-charge plan would almost certainly pass the first two tests, but that the third may be a difficult obstacle if substantial revenues are collected.

In *American Airlines, Inc.* v. *Massachusetts Port Authority*, several airlines challenged landing fees at Logan Airport on the grounds that the charges included costs of projects from which they received no benefit (an unused runway, landfill, and payments to a nearby hospital). The Court of Appeals rejected the challenge, stating

This is not to say that states can run wild and tax users for all extravagances. The facilities must be relevant to the operation of the airport. And the revenue from the landing fee must be fairly consonant with the costs incurred. But within those broad parameters users share both the benefits and costs of an airport's decision.

It is possible to argue that such challenges are not relevant, because they involve user fees intended to raise revenues whereas revenues are an incidental byproduct of a noise charge (Nierenberg 1978).

Noise charges also must not place an undue burden on interstate commerce. Since noise charges would interfere less with interstate commerce than a nighttime curfew would, and the courts have upheld airport proprietors' curfews, Nierenberg (1978) reasons that noise charges are constitutionally permissible. In *National Aviation* v. *City of Hayward*, the court found that the effect of a curfew was "incidental at most and clearly not excessive when weighed against the legitimte and concededly laudable goal of controlling the noise levels . . . during late evening and morning hours." This argument, however, seems to beg some important questions: Would the court's conclusion have been the same if the curfew had been on activity at O'Hare or Kennedy Airport? Would a noise charge that made many night operations (especially freight operations) at a major airport uneconomical be an excessive burden on interstate commerce?

To avoid conflict with federal law, noise charges must be judged "reasonable" and "nondiscriminatory." Both Nierenberg (1978) and Rosenthal (in Massport Master Plan Study Team 1975) conclude that noise charges would be legally acceptable. Rosenthal states

Federal legislation requiring that landing charges be reasonable would not be violated by charges that reflected the rational purpose of encouraging the use of quieter aircraft. The requirement in air terminal leases and in FAA airport improvement grants that landing charges be nondiscriminatory would also seem not to be violated, since a fair interpretation of that term would require only that there be no discrimination between airline and airline, not between plane and plane, charges scaled to decibels would seemingly be just as valid as charges scaled solely to aircraft weight.

Nierenberg (1978) also contends that noise charges would be consistent with state laws and procedures.

Contracts between the airport and the airlines present perhaps the toughest legal obstacle to the enactment of noise charges. As mentioned, most airports are financed by airline-backed revenue bonds. The contracts relating to this financing typically specify the method for calculating landing fees. Some contracts even block renegotiation of landing fees for as long as 30 years. Current expiration dates of the contracts at the major U.S. airports range from the annual negotiation permitted yearly at Midway to the year 2009 for Detroit (Miller 1979, p. 154). Nierenberg (1978) suggests that, because many of the airport-airline contracts were entered into before aircraft noise became a serious problem, the common-law doctrine of changed circumstances may render the landing-fee provisions unenforceable. But the courts could force airports to postpone changing the basis for landing fees until the contracts were renegotiated.

In summary: It is clear that an airport operator wishing to implement a system of noise charges faces difficult legal obstacles. If existing airline-airport contracts are upheld, it will be many years before noise charges can be introduced at many major airports. These legal obstacles obviously reduce the appeal of noise charges to airport operators.

Chapter 5

An Evaluation of Incentive-Based Strategies

The current interest in regulatory reform suggests that economists' pleas for economic incentives may receive closer attention. Arguments in support of such schemes will be most persuasive if proponents acknowledge the limitations of economic-incentive approaches and apply them to the most appropriate cases. In addition, analyses supporting the use of economic-incentive schemes should go beyond comparisons of first-order costs and benefits to include administrative obstacles and other considerations.

The purpose of this case study has not been to present policy recommendations for regulating aircraft noise (which would quicky become obsolete), but rather to shed light on the factors that make economic-incentive schemes attractive. Although the following conclusions relate specifically to aircraft noise control, they should thus be viewed as part of a more general investigation of economic-incentive strategies.

The charge approach appears attractive as a national strategy for controlling aircraft noise.

Information is available to enable the regulator to set a charge that approximates the marginal damage caused by individual aircraft operations. Because noise is not primarily a health problem, these valuations do not raise the philosophical and ethical controversy that surrounds charges for most air or water pollutants. The damage-based noise charge is therefore likely to be more readily acceptable than a charge on pollutants that cause death or illness, or on pollutants for which data with which to estimate damage functions with reasonable accuracy do not exist.

Cost savings with a noise charge are likely to be substantial. The costs of reducing noise damage differ widely for different aircraft, and the regulator would find it difficult to determine these differences. Decentralizing decisions on noise abatement to the airlines would thus increase the cost-effectiveness of control. Allowing airlines to pay a charge when control is expensive avoids the bluntness of noise standards for all aircraft.

Information on aircraft emissions is inexpensive to obtain and verify. Monitoring difficulties are often major impediments to effluent-charge schemes, because levying the charge requires accurate data on the quantities of pollutants emitted. Monitoring aircraft noise is relatively easy, partly because the number of airlines is small and aircraft operations are highly visible. Airlines currently submit information on their takeoffs and landings by airport; these data are already sufficient for a simple charge scheme based on prototype noise levels. Relatively inexpensive monitoring of individual flights could be carried out at major airports. The use of monitored data would increase the cost-effectiveness of the program by encouraging airlines to change some noisy landing and takeoff procedures, and would also settle any disputes about the accuracy of the prototype values.

The enforcement of noise standards is inevitably characterized by brinksmanship, confrontation, and delay. The recent congressional modification of the retrofit rule clearly demonstrates the limitations of this approach. Each airline was originally required to meet noise standards for a portion of its fleet by 1981 and for all of its fleet by 1985. The airlines argued that meeting this timetable would be costly and might force the discontinuance of service to small communities rather than the retrofitting of two- and three-engine aircraft. The airlines also wanted the federal government to pay for the noise-control modifications. Although noncompliance would in theory result in aircraft being grounded, the FAA was unlikely to impose such a drastic penalty, thus allowing the airlines significant bargaining power after the regulation was in place. In a confrontation that lasted almost two years, Congress rejected plans to pay for the modifications but did extend the compliance deadlines for airlines operating the two- and three-engine aircraft. The major criticism of this process is that delaying the requirement for the four-engine aircraft (707s and DC-8s) would have been more beneficial because they are the most costly to retrofit. Of course, it is not clear that a noise charge would not elicit confrontation and modification; it would, however, have made Congress less likely to pass up inexpensive opportunities to reduce aircraft noise and retain the more costly ones.

The noise charge promises to speed up the introduction of a new generation of quieter aircraft. One of the traditional arguments for charges is that they provide incentives to reduce emissions below the current standards. It seems unlikely, though, that a noise charge would induce the airlines to spend more on noise-control technology for their current aircraft. By permitting airlines to avoid costly modifications of some existing aircraft, however, the noise-charge scheme might hasten the introduction of quieter and more fuel-efficient "third-generation" planes into their fleets.

The noise-standard approach can be modified to reduce its disadvantages with respect to the noise-charge scheme.

There is a danger of characterizing regulation as "straw men"—senseless restrictions either enforced inflexibly by government bureaucrats or manipulated by the companies whose emissions are regulated. It is clear, however, that noise-control regulations can be made more flexible and thus made to approximate many of the strengths of the charge scheme. For example, adding a set of penalties related to the gains from noncompliance would relax some of the inflexibility of regulations and reduce the brinksmanship that it generates. Moreover, phasing in the regulations could avoid very high marginal costs of control. The greatest expense of compliance would probably be the retrofitting of the oldest and noisiest 707s and DC-8s—precisely the costs that a phased compliance schedule avoids. There is no guarantee that adhering to the phased timetable would avoid all high-cost compliance.

Virtually all incentives the airlines have to delay enforcement would be eliminated if the federal government subsidized the retrofitting or the replacement of noisy planes. Programs of this sort tend to generate high costs, however, because the grants are not related to specific noise reductions. Retrofitting the entire existing fleet of aircraft is not desirable on cost-benefit grounds.

Permitting regulations to vary by airport would probably increase markedly the cost-effectiveness of noise control.

Like all federal environmental controls, the current noise regulations apply equally throughout the country even though the noise problem varies enormously from one airport to the next. (More precisely, the marginal benefits of noise control vary much more than the marginal costs.) Aircraft noise regulation can thus be made more cost-effective by exploiting the differences in control benefits, requiring stringent controls where benefits are great and lenient controls where they are small. Our estimates suggest that stricter compliance timetables for large- and moderate-benefit airports would reduce the costs of noise control substantially with very little sacrifice in benefits. The estimated net benefits of switching to a geographically varying strategy range from $302.4 million to $412.9 million.

It is impossible to predict the precise net benefits of using a varying charge, but it is likely that the advantages of the charge approach increase when geographic variation is allowed. Providing different regulations for different airports is a relatively clumsy process. The noise charge, however, could be calculated separately for each airport to approximate the marginal damage of aircraft operations. (The noise-charge formula discussed here also varies by decibel level and time of day.) Airlines could then allocate their planes and make decisions on retrofitting or replacement on the basis of more precise damage estimates.

Although any regulatory strategy would have a differential impact on airlines, the variations would probably be smaller under the noise charge than under the noise standard.

Airlines operating the older and more expensive to retrofit 707s and DC-8s would incur greater costs than airlines with newer aircraft that already meet the Part 36 standards. Since the net burden of retrofitting would also depend upon fare increases (which are impossible to predict in detail), it is very likely that some airlines' fare increases would be greater than their cost. A noise charge assessed on all decibels above a certain threshold would decrease this variation in costs among the airlines because it would raise the floor and lower the ceiling for cost burdens. Airlines operating quiet aircraft that incurred little or no cost under the noise standards would pay some charge, while airlines with very noisy aircraft would be able to pay the charge and avoid very high compliance costs. The airlines may thus view the noise charge as more equitable.

Switching from a noise standard to a noise charge would, however, probably increase the airlines' total costs. Although they could eliminate costs by paying a charge in some cases, the airlines would have to pay a charge on noise above the threshold. The amount of revenue collected could be reduced by changing the threshold, adopting a proportional damage charge rather than a full damage charge, or creating a mixed scheme of subsidies for "low-noise" operations and charges for "high-noise" operations.

The noise charge is a less attractive option from the perspective of the individual airport operator and therefore is not likely to be adopted.

Although a noise charge adopted by a single airport would have advantages when judged from a national perspective (cost-effectiveness, closeness to an optimal level of aircraft noise, and encouragement of airline competition), it entails many disadvantages for the airport operator. The major drawback of the noise charge is that it creates major uncertainties about the level of noise control to be achieved and the means of reducing aircraft noise. The airport operator is likely to be particularly concerned about reductions in flights that might jeopardize the airport's role in the local or regional economy.

The local noise-charge scheme has other drawbacks. It would change the revenues resulting from landing fees and their allocation among the airlines—changes that the airport operator would probably perceive as unfair. State and local government might welcome increased revenues, but airport officials might see revenue potential as a threat to their autonomy. At the very least, the increased revenue would subject an airport to intense opposition from the airlines. Though a noise-charge scheme could be quite simple and inexpensive to administer (even with monitoring), objections from airlines over the accuracy or the legality of the charges could mire the program in administrative confusion. In addition, existing airport-airline contracts might prevent the airports from legally adopting a noise-charge scheme. Without a major federal or state initiative, an individual airport is therefore unlikely to adopt a charge scheme.

Although the noise-charge scheme would generate revenue for the public, it is not clear that the money would be well spent or welcomed by the FAA or airport operators.

Raising revenue by taxing socially harmful activities is generally good public policy. The substitution of noise-charge revenue might reduce inefficiencies generated under the other revenue sources. The payments might also be considered fair because they represent a "user tax" for the airlines' right to use the peace and quiet around the airport. There is a danger, however, that the revenues would be employed for relatively inefficient projects. There might be pressure to use the money to sound-proof buildings, compensate noise-impacted persons, or purchase noise-impacted properties—measures that may not be cost-effective means of reducing noise damage.

Creating a fund from the noise-charge revenue would change the traditional objectives of the FAA and the individual airports, who may not want to take on the additional tasks of deciding on appropriate compensation or public noise-reduction projects. Airport operators in particular have viewed landing charges as a means of recovering direct airport costs not covered by other revenue sources, and are likely to be hostile to imposing charges that exceed these costs. Indeed, there may be substantial legal obstacles to generating such excessive revenues. Of course, local governments might find it attractive to tap an additional revenue source, especially one whose burdens would be borne by persons outside the jurisdiction; airport operators, though, would likely resist any reduction in their autonomy implied by their becoming a major revenue source.

The marketable-rights scheme appears to be a less attractive option than the noise-charge scheme at the national level, but might be more attractive for an individual airport.

The arguments in favor of setting a noise charge—that the benefits of noise reduction are linear and reasonably well known, whereas the costs are much less well known and probably nonlinear—imply that the establishment of a particular number of noise rights would probably be arbitrary. Though the marketable-rights scheme may share many of the advantages of the charge scheme (incentives for technological development, minimum cost of achieving the given level of noise reduction, and less potential for brinksmanship), the arbitrariness of the allowable noise levels makes the scheme less attractive on overall efficiency grounds.

In addition, the administrative costs of establishing a noise-rights program might be substantial. No empirical evidence exists on these matters, but it might be difficult and costly to arrange an orderly market with a relatively stable price or with future markets to allow airlines to plan for noise reduction. The marketable-rights system might also permit the airlines to engage in anticompetitive, predatory behavior. Most emission-rights proposals assume that some companies do not participate in both the rights market and the product market. Aircraft noise is only emitted by a small number of airlines, however, and the number is even smaller at any particular airport. The loss of market control brought about by airline deregulation may lead some airlines to attempt to find new means of blunting the effects of competition, and limiting new entrants by strategically intervening in the noise-rights market might be one avenue. Airlines might also collude to produce a high market price for noise permits in the expectation that the government would then add to the number of noise rights.

A marketable-rights scheme might have many more attractions for an individual airport operator, because it would reduce some of the uncertainties associated with the noise charge. Establishing rights for noise in general (for example, total number of decibels) would have the advantage of establishing a limit, but would create uncertainties about how that limit would be achieved; the airlines might reduce service rather than substitute quiet flights or switch to daytime operations. The airport would therefore probably favor a variant of the marketable-rights scheme that would set the method of noise reduction as well as the level. For example, the airport might establish rights to a number of night operations, or to a number of night operations in noisy aircraft.

Case Study 2

The Regulation of Airborne Benzene
Albert L. Nichols

Chapter 6

The Problem of Benzene

In June 1977 the Environmental Protection Agency listed benzene as a "hazardous air pollutant" under section 112 of the Clean Air Act, primarily because of evidence that exposure increases the risk of leukemia (42 *Fed. Reg.* 29332, 1977). After that listing, in preparing regulations, the agency conducted or sponsored numerous studies of health effects, major sources of exposure, and techniques for controlling benzene emissions from various sources. In April 1980 the agency formally proposed the first in a planned series of "benzene regulations," a standard that would limit emissions from plants using benzene as a feedstock to produce maleic anhydride (45 *Fed. Reg.* 26660, 1980).[1]

Benzene is a major industrial chemical used primarily in the production of other industrial chemicals, which in turn are used to produce a variety of widely used substances, including polystyrene, polyester, nylon, and polyurethane foam (Mara and Lee 1978, p. 21). In production volume, benzene ranks among the top fifteen industrial chemicals (*Chemical and Engineering News*, June 9, 1980, p. 36). Benzene is also a constituent of gasoline and other petroleum products; it is a natural component of crude oil, and additional amounts are formed during the refining process.

Benzene also has been the target of regulatory action by the Occupational Safety and Health Administration, which promulgated a new standard in February 1978 that lowered permissible occupational exposure limits from 10 parts per million to 1 ppm (43 *Fed. Reg.* 5918). The Fifth Circuit Court of Appeals struck down that standard in October 1978, primarily on the grounds that OSHA had failed to demonstrate that the new standard would yield significant benefits (581 F.2d 493). The Supreme Court upheld the lower court's ruling, though on somewhat narrower grounds (100 S.Ct. 2844, 1980).

The potential costs of regulating benzene are great. OSHA estimated that its own standard would impose capital costs of $267 million, with annual operating costs of $124 million in the first year and $74 million thereafter (Arthur D. Little and Co., 1977, p. 1-8). How costly the

EPA's regulations will be depends on what activities it decides to regulate and what specific standards it requires. If the EPA chooses some of the more extreme steps, such as requiring vapor recovery at service stations, the costs could run to billions of dollars.

The EPA's approach to benzene is important in its own right, but also as a prototype for other environmental carcinogens. Many actions taken with regard to benzene are likely to set important precedents. Over the next few years the EPA will almost certainly develop regulations for other carcinogens. In October 1979 it issued a "notice of proposed rulemaking" describing a "generic" policy for identifying and regulating airborne carcinogens (44 *Fed. Reg.* 58642). The EPA's own view of benzene as an important precedent for this policy is indicated by the fact that, when asked for a case study to illustrate how the policy might be implemented, it recommended benzene (Regulatory Analysis Review Group 1980). Benzene also has played a key role in OSHA's developing approach to the regulation of carcinogens. Many provisions of OSHA's generic policy on occupational carcinogens, promulgated in January 1980, are virtually identical to that agency's benzene standard.

Benzene's potential role as a prototype for future regulation makes it a particularly attractive vehicle for studying alternative forms of regulation. This attractiveness is enhanced by the fact that benzene is emitted from a variety of sources, so that in a single case study it is possible to deal with a range of different regulatory problems requiring different regulatory strategies.

This study is designed to be of interest to officials concerned with benzene regulation *per se*. More specifically, it examines the possibility of using incentive mechanisms or of modifying traditional emission standards to achieve more efficient outcomes. It also offers guidance concerning the applicability of different strategies to the regulation of environmental carcinogens, of which benzene is a representative.

Health Effects Associated with Benzene

Benzene has long been recognized as hazardous. Reports of toxic effects in humans and laboratory animals date back to the last century (U.S. Environmental Protection Agency 1978a, p. 29). Until recently, however, regulatory efforts focused on preventing noncarcinogenic effects, both acute and chronic, associated with exposure at high concentrations rarely found outside the workplace. Thus, at present the EPA has no regulations directed specifically at benzene, though some benzene emissions are regulated under standards covering hydrocarbons as a class.

Recent actions by the EPA and OSHA have been spurred by mounting evidence that benzene is a leukemogen. Many scientists believe that any exposure to a carcinogen or a leukemogen poses some risk—that, unlike other toxic effects, cancer may be induced by even minute doses, so that we cannot draw great comfort from the fact that ambient exposures to benzene generally are several orders of magnitude lower than the levels at which harmful effects have been confirmed in workers. The EPA and OSHA take the position that there is no safe level of exposure to a carcinogen. In addition to leukemia, the EPA has identified pancytopenias (including aplastic anemia) and chromosomal aberrations as potential concerns at low levels of exposure (U.S. Environmental Protection Agency 1978a).

Leukemia
Dozens of reports linking leukemia with benzene exposure have appeared in the medical literature since 1928 (U.S. Environmental Protection Agency 1978a, p. 68). Until several years ago, however, the evidence was considered inconclusive because it consisted almost entirely of clinical reports of isolated cases of leukemia in workers exposed to high concentrations of benzene and, often, other potentially carcinogenic chemicals. As recently as 1976, a review by the National Academy of Sciences concluded that "it is probable that all cases reported as 'leukemia associated with benzene exposure' have resulted from exposure to rather high concentrations of benzene and other chemicals" (National Academy of Sciences 1976, p. ii). The inability of investigators to induce an abnormal incidence of leukemia in laboratory animals through exposure to benzene and the negative results of several epidemiological studies contributed further to skepticism about the leukemogenicity of benzene.

Current regulatory attention to benzene was prompted in large part by the completion, in early 1977, of an epidemiological study of workers who were exposed to solvents containing benzene while they were employed at two plants in the rubber industry. That investigation showed a much higher than normal incidence of leukemia among the 748 workers studied (Infante et al. 1977). The results led OSHA to promulgate its emergancy temporary standard in May 1977, and to propose its new permanent standard, which lowered permissible occupational exposures tenfold. Presumably, that study also played a major role, together with OSHA's actions, in the EPA's decision to list benzene as a hazardous air pollutant in June 1977 (in response to a petition from the Environmental Defense Fund).

Several additional epidemiological studies provide mixed support for the results of Infante et al. Thorpe (1974) studied the health status of 36,000 employees and annuitants of European affiliates of a major oil company. For the overall group, he found no statistically significant increase in the incidence of leukemia, though there was, in the words of an EPA review, "a tendency toward higher leukemia rate in the benzene-exposed as compared to the unexposed work groups" (U.S. Environmental Protection Agency 1978a, p. 79). Similarly, Redmond et al. (1976) found no evidence of a higher leukemia risk among coke-oven workers, even among those working in byproduct departments where benzene is produced.

Ott et al. (1977) studied 594 workers exposed to benzene in the chemical industry over several decades. They discovered a higher-than-expected incidence (3 leukemia deaths versus 0.8 expected), but the results were only of borderline statistical significance. Moreover, those results have been questioned because some of the workers involved, including those who died of leukemia, were also exposed to other suspect chemicals. Nevertheless, the Ott study is of interest because some effort was made to estimate exposure levels for different groups of workers and because the exposures appear to have been relatively low (U.S. Environmental Protection Agency 1978a, p. 83).

In contrast, reports of benzene-induced leukemia among Turkish shoe workers show a much stronger effect. The exposure data are almost nonexistent, but what data there are suggest very high exposures, on the order of hundreds of parts per million (Aksoy et al. 1974, 1976, 1977). These studies are particularly interesting because they show a sharp rise in pancytopenias and (with a greater lag) leukemia after the introduction of adhesives with a high benzene content. Moreover, their results show a reduction in those two types of disease as benzene-based adhesives were phased out of the Turkish shoe industry (U.S. Environmental Protection Agency 1978a, p. 72). Similarly, Vigliani et al. reported that the incidences of pancytopenias and leukemia rose sharply in the Italian rotogravure industry after inks and solvents containing large amounts of benzene were introduced (U.S. Environmental Protection Agency 1978a, p. 74).

The epidemiological evidence that benzene is a leukemogen appears to be very strong. The implications of that evidence for low-level environmental exposures are much less clear, however. The epidemiological data are all drawn from occupational settings where exposures ranged from several parts per million to several hundred on a regular basis, with peaks exceeding 1,000 ppm. In contrast, typical environmental exposure levels average several parts per billion (ppb)—at least 1,000 times lower. As noted earlier, many scientists agree with the official EPA position that the only safe level of exposure to a carcinogen is no exposure. Others, however, believe that chemical carcinogens exhibit thresholds—doses below which no risk is incurred. If that view is correct, current environmental exposures to benzene are almost certainly harmless.

With the current knowledge about carcinogenesis, the controversy over the risks posed by low-level exposures cannot be resolved conclusively. The EPA's health-risk assessment on benzene draws the following conclusion:

Unfortunately, the data are not adequate for deriving a scientifically valid dose-response curve. Such a curve may be estimated on the basis of various assumptions; these assumptions, however, usually represent hypotheses that, although they may be valid, are not yet proven (U.S. Environmental Protection Agency 1978a, p. 94).

The EPA's Carcinogen Assessmment Group, however, has derived a dose-response function based on the assumption that risk is proportional to cumulative exposure. Using data from the epidemiological studies by Infante et al., Askoy et al., and Ott et al., the CAG estimates a risk of 0.339×10^{-6} deaths per ppb-person-year (Albert et al. 1979). According to this estimate, exposing 10 million people to one ppb of benzene for one year would result, on average, in 3.4 extra deaths from leukemia. This particular estimate and the general problem of estimating dose-response functions are addressed in detail in chapter 7; at this point, the reader should note that the linear dose-response model yields much higher risk estimates at low exposure levels than other extrapolative models.

Pancytopenia

The term *pancytopenia* refers to a variety of blood disorders, including aplastic anemia. In milder cases, patients may complain of "lassitude, tiredness, easy fatigability, malaise, dizziness, headaches, palpitation, and shortness of breath" (U.S. Environmental Protection Agency 1978a, p. 52). In extreme cases, individuals with pancytopenia may die from hemorrhaging or from severe infections resulting from a reduction of those elements in the blood that fight infection. Some patients with pancytopenia later develop leukemia (U.S. Environmental Protection Agency 1978a, p. 50).

The evidence that exposure to benzene causes pancytopenia and other disorders of the blood is both long-standing and strong. The effects, however, are restricted to workers exposed at high levels; those exposed at lower levels may develop much less severe, nonfatal disorders, such as ordinary anemia. It seems highly unlikely that even these milder disorders would be induced in populations exposed to ambient benzene, since the levels of exposure are so much lower.

Chromosomal aberrations
Laboratory experiments with animals and human epidemiological studies indicate that exposure to benzene can cause chromosomal damage. The EPA's review of the health effects of benzene concludes that "the available documentation strongly suggests that chromosomal breakage and rearrangement can result from exposure to benzene and that damage may persist in hematopoietic and lymphoid cells" (U.S. Environmental Protection Agency 1978a, p. 21).

The possibility of chromosomal aberrations raises two concerns: To the extent that the effects are permanent and heritable, birth defects may result; and many scientists believe that mutagenic substances are also likely to be carcinogenic, so that evidence of chromosomal damage supports the hypothesis that benzene is a leukemogen. The EPA's review emphasizes the latter concern; benzene has not been associated with birth defects, nor do the data permit even a crude estimate of the relationship between exposure and the frequency of chromosomal abnormalities. Thus, it appears that chromosomal damage mainly strengthens the case for benzene-induced leukemia.

Production and Use of Benzene

Nearly all benzene produced in the United States is derived from petroleum at refineries or associated petrochemical complexes. Coke ovens, where benzene is a byproduct, account for only about 5–8 percent (PEDCo Environmental, Inc., 1977; p. 4-18). In 1979, U.S. production was about 12.7 billion pounds (about 1.7 billion gallons), up 16.3 percent from 1978 (*Chemical and Engineering News*, June 9, 1980, p. 36).

Benzene constitutes only about 0.15 percent of crude oil, but additional benzene is formed during refining, so that U.S. gasoline contained, on average, about 1.3 percent benzene in 1977, with a reported range of 0.2–4.0 percent (Turner et al. 1978, p. 1-5). Since gasoline consumption is about 100 billion gallons per year (Mara and Lee 1978, p. 112), the benzene in gasoline is almost equal to that produced for nonfuel use.

About 97 percent of the benzene consumed in nonfuel uses is employed as an intermediate in the production of other industrial chemicals (Mara and Lee 1978, p. 65). Ethylbenzene alone accounts for about 50 percent; cumene and cyclohexane each consume about 15 percent of benzene output (*Chemical and Engineering News*, November 19, 1979, p. 20). Benzene also is used to produce chlorobenzene, detergent alkylate, maleic anhydride, and nitrobenzene. These chemicals are used in turn to produce a variety of products, such as polyurethane foams, nylon fibers, insecticides, and reinforced plastics (Mara and Lee 1978, p. 21). Minor amounts are used in some solvents and adhesives, though such uses have declined as awareness of the hazards associated with benzene has increased.

Emissions and Exposure

The EPA has commissioned several studies to identify major sources of benzene emissions and exposure. The discussion below focuses on the two studies—one on emissions, the other on exposure—that attempt to cover the full range of source categories. These studies are crude and are subject to major uncertainties, but they provide some feel for the problem and an indication of the relative importance of different sources. Several studies of specific source categories, such as maleic anhydride plants and service stations, are discussed later in connection with regulatory alternatives for those sources.

Emissions
Estimating emissions from different source categories poses several difficulties. Ideally, the estimates would be based on direct measurements from individual sources, but that would be excessively expensive given the large number of sources. Instead, the EPA's contractor, PEDCo Environmental, Inc., used a variety of indirect ways to derive its estimates. In a few cases, estimates of nationwide emissions from particular source categories were already available, and PEDCo merely reported those estimates without indicating how they were developed. In most

Table 6.1   Estimated benzene emissions.

	Millions of pounds	Percent of total
Chemical manufacturing	60.0	11.0
Coke ovens	7.8	1.4
Petroleum refineries	4.1	0.8
Solvent operations	N.A.[a]	—
Storage and distribution of gasoline and benzene	10.2	1.9
Auto emissions	443.6	81.5
Service stations	14.6	2.7
Other miscellaneous	4.0	0.7
Total assessed	544.3	

Source: PEDCo Environmental, Inc., 1977, pp. 1–2, 4–62.
a. Not assessed.

cases, however, PEDCo combined estimated emission factors with data on production capacity to estimate total emissions.

Table 6.1 reports PEDCo's estimated emission levels. Automobile emissions account for over 80 percent. Chemical manufacturing is the next largest category, with 11 percent; over half of that is from maleic anhydride plants. Service stations contribute 2.7 percent.

The accuracy of table 6.1, however, is questionable. For each source category, PEDCo rated the estimate of emissions as A ("good emission data available"), B ("fair estimate ± 100%"), or C ("poor data base, order of magnitude estimate") (PEDCo Environmental, Inc., 1977, p. 4-2). Only one estimate, that for maleic anhydride plants, rated an A. If, however, we compare the PEDCo estimate of emissions from maleic anhydride plants with the estimate presented in the draft environmental-impact statement for such plants, which was based on a much more detailed and presumably accurate study, the difference is startling. PEDCo estimated that the emissions level was 34.8 million pounds, or about 15.8 million kilograms. The draft EIS, in contrast, estimated that annual emissions would total only 5.8 million kilograms at full-capacity operation (U.S. Environmental Protection Agency 1980, p. 1-13). This 63 percent reduction reflects a lower estimated uncontrolled emission rate and the fact that over half of the benzene-fed maleic anhydride plants already have some type of emission-control device; PEDCo assumed that no plants had emission controls.

If we scale down the EIS emission estimate to take account of less-than-full-capacity operation, the differences are still more dramatic.

For example, in 1977 maleic anhydride plants operated at about 56 percent of nominal capacity (U.S. Environmental Protection Agency 1980, p. 1-1). If we assume that emissions were reduced proportionately, estimated emissions in 1977 would have been only 3.3 million kilograms, little more than one-fifth the level estimated by PEDCo. Similarly, though less dramatically, whereas PEDCo estimated benzene emissions from the production of ethylbenzene and styrene at 4.2 million kg/ year, the draft EIS for ethylbenzene/styrene plants puts current annual emissions at 2.2 million kg, given "present control levels and present production levels" (U.S. Environmental Protection Agency 1979, p. 1-5).

Exposure
The EPA followed up the PEDCo study of emissions with an assessment of exposure levels conducted by SRI International (Mara and Lee 1977, 1978). That study combined emission estimates with crude dispersion modeling and population data to estimate the numbers of people exposed to various concentrations of benzene from different sources. The results also were used to generate estimated levels of total annual exposure, measured in ppb-person-years, for each type of source. The results of the final study are shown in table 6.2.[2]

As with emissions, automobiles account for over 80 percent of estimated exposures. Those exposures, however, should fall rapidly over the next several years as the fraction of automobiles with catalytic converters increases. Although current regulations do not restrict benzene emissions *per se*, they do limit hydrocarbons as a class. Under existing regulations, PEDCo estimates that by 1985 benzene emissions from automobiles will fall by 76 percent from the level shown in table 6.1 (PEDCo Environmental, Inc., 1977, p. 1-5). If exposures fall by roughly the same proportion, the potential health effects associated with benzene emissions from automobiles will also be reduced fourfold.[3]

These exposure estimates should be viewed as giving only a rough indication of relative magnitudes. Several factors suggest that little faith should be placed in their precision. First, for the most part they incorporate the PEDCo emission estimates, with their associated errors. Second, the dispersion modeling was very crude and subject to considerable uncertainty. Third, several implausible assumptions were made regarding such variables as plant locations, capacity utilization rates, and the degree to which emissions already are controlled. Finally, Mara and Lee's study is marred by several inconsistencies and simple com-

Table 6.2 Estimated ambient exposure to benzene.

	Millions of ppb-years	Percent of total
Chemical manufacturing	8.5	4.7
Coke ovens	0.2	0.1
Petroleum refineries	2.5	1.4
Solvent operations	N.A.[a]	—
Storage and distribution of gasoline and benzene	N.A.	—
Auto emissions	150.0[b]	82.5
Gas stations		
Nearby residents	19.0[c]	10.5
Self-service users	1.6[d]	0.9
Other miscellaneous	N.A.	—
Total assessed	181.8	

Source: Mara and Lee 1978, p. 3.
a. Not assessed.
b. Applies only to residents in standard metropolitan statistical areas with populations greater than 500,000.
c. Applies only to "urban" residents.
d. Applies only to individuals filling their own gas tanks at self-service stations.

putational errors that raise the possibility that a more comprehensive and detailed review might uncover additional serious problems. Some illustrative examples follow.

As shown in table 6.2, Mara and Lee estimated that automobiles account for 150 million ppb-person-years of exposure. An earlier version of the study put the total at 102.2 million ppb-person-years, based on exactly the same dispersion modeling and population estimates. The difference seems puzzling, but resulted merely from a change in the way the data were grouped and averaged. My examination of the dispersion-modeling results suggests that the earlier, lower figure reflects the data more accurately.[4]

The problems with the service-station estimates are still more serious. Mara and Lee's (1977, p. 4) original estimate was 90 million ppb-person-years, based on dispersion modeling for a "typical" station. The "typical" station, however, was four times larger than the national average, and the conversion factor used to derive average concentrations from eight-hour worst-case dispersion modeling was 2.5 times too high. The EPA then used a different dispersion model that yielded an estimate of 1.2 million ppb-person-years—lower by a factor of 75.[5] The final estimate of 19.0 million ppb-person-years did not rely on the dispersion model's results, but rather was based on a distance-concentration curve estimated using only 15 measurements from each of two service stations, with no controls for pumping volumes, wind speed, or other important factors.[6] Even more remarkable, the estimate of exposure for self-service gasoline customers is based on an average of measurements for three customers at a single station on a single day. Thus, it requires an act of faith to attach much credence to the exposure estimates for service stations.

The exposure estimates for chemical manufacturing also must be treated skeptically. They incorporate the errors in emission estimates noted earlier, and are based on full-capacity operation and center-city locations for all plants located in urban areas. Both assumptions increase estimated exposures.

Estimated overall risk
The list of problems discussed above is undoubtedly incomplete. It appears that in most cases where a range of plausible assumptions existed, the EPA and its contractors systematically selected those that resulted in higher estimates of emissions, exposure, and risk. Thus, although the crudeness of the estimates precludes any definitive conclusions, it seems likely that the figures presented are too high.

Even if we take the EPA's estimates at face value, benzene in the environment does not appear to be a major threat to health. A crude estimate of overall risk from the emissions identified by the EPA can be obtained by combining Mara and Lee's estimate of total exposure with the Carcinogen Assessment Group's estimated dose-response function:

$(0.339 \times 10^{-6}$ deaths/ppb-year$)(181.8 \times 10^{6}$ ppb-years/year$)$

$= 61.6$ deaths/year.

This estimate is almost certainly too high, both because the exposure estimates are probably excessive and because (as discussed at greater length in the next chapter) the CAG's dose-response function probably overstates the risk.

Regulatory Options

In view of the apparently small impact on public health of environmental benzene and benzene's many important uses, a complete ban is not warranted. Some sources, however, may offer cost-effective opportunities for control. We are faced, then, with the questions of which categories to regulate (if any) and how to regulate them. This study focuses on the appropriate targets and instruments, though the question of whether any regulation makes sense for some sources also will be addressed. The desirability of regulation, and its optimal form, will almost certainly vary across sources. A single, uniform approach to regulating benzene is likely to be highly inefficient.

The constraints placed on the EPA by existing statutes, specifically the Clean Air Act, have not been considered in formulating and assessing the regulatory options discussed here and in later chapters. Many of the alternatives analyzed could not be implemented under existing laws, but rather would require new legislation expanding the tools available to the EPA and altering some of the decision criteria to be used in setting regulations.

Alternative instruments
In addition to standards, three incentive-based instruments might be considered: charges, marketable permits, and subsidies.

Subsidies suffer from several faults. In particular, they distort the prices of final goods and may actually increase production of goods associated with negative externalities. Subsidizing the reduction of benzene emissions from maleic anhydride plants, for example, might lower the cost of producing maleic anhydride, and hence its price, thus artificially stimulating consumption and production. In addition, subsidies require government expenditures, necessitating cutbacks in other programs or the imposition of additional taxes. These factors and others suggest that subsidies will generally be less desirable than charges or marketable permits.[7]

The choice between charges and marketable permits is less clear-cut. The two instruments yield equivalent results under conditions of certainty. However, when control costs and damages are uncertain, charges are superior to marketable permits if marginal damages do not vary greatly with the level of emissions and exposure. For benzene and other carcinogens, the EPA has adopted a linear dose-response function (constant marginal damage) as its working hypothesis. Although that assumption may be pessimistic, with one of the less conservative models the estimated risks would be infinitesimal. Thus, a decision to regulate presupposes acceptance of the linear dose-response model, or at least a belief that there is a reasonable chance that it is correct. In either case, the expected marginal damage is constant and charges are more efficient than permits.[8]

Marketable permits also pose problems in defining market boundaries. The rationale behind marketable permits is the importance of achieving a target level of emissions or exposure. The effects of benzene emissions are quite localized, however, so it would not make sense to allow maleic anhydride plants in different regions (for example) to trade permits with one another; their emissions do not affect the same individuals. For other sources, such as service stations, the effects are still more localized. Thus, a single national market in benzene emission permits, or even regional markets, would not ensure uniform exposure levels. Only a system with dozens or hundreds of separate markets could achieve that goal. Such a system would be difficult to administer, and most of the individual markets would be too "thin"; potential purchasers would be too few to sustain competition, and many of the desirable properties of such markets would be lost.[9]

Given the problems with both permits and subsidies, this analysis will focus on charges as the alternative to standards. Many of the features ascribed to charges, however, will hold for marketable permits and (less often) for subsidies as well.

Alternative targets
Regulators must choose the target as well as the instrument. Emissions, the target of most EPA standards, are not always the optimal target. The analysis of maleic anhydride plants in chapter 8 shows that when the link between emissions and exposure or damage varies across sources, a uniform charge on emissions will not be optimal. The regulator may increase efficiency by altering the target of regulation from emissions to exposure or damage, whether the instrument is a charge or a standard.[10] In other cases, emissions may not even be a feasible target. Service-station emissions, for example, cannot be monitored continuously. A standard or a charge may have to be based on the type of vapor-recovery equipment installed, rather than directly on emissions. Similar problems arise with "fugitive" emissions, such as valve and pipe leaks in refineries and chemical plants, where the number and small size of potential sources make direct monitoring prohibitively expensive. In such cases, the target may be maintenance and operating procedures.

Chapter 7

Estimating the Damage Function

To analyze the efficiency of alternative strategies for benzene, we need to estimate the damages caused by exposure. The need for such an estimate is perhaps most obvious with regard to damage-based charge systems, where the level of the charge is tied explicitly to damages. However, the efficiency of standards and other quantity-based approaches is equally sensitive to the accuracy of the estimated damage function. The problem is unavoidable; if we wish to regulate on an efficient, rational basis, some attempt must be made to estimate the damages. Estimated damages are the only basis for predicting the benefits of standards and alternative regulatory strategies.

Estimation of damages can be broken down into two components: the physical consequences (such as leukemia deaths) of exposure, and the dollar value placed on the consequences. Neither can be estimated with much precision. Scientists disagree about the risks of low-level exposures, and there is no widely accepted method for valuing the benefits of increased health and longevity. We can, however, make crude estimates and place rough boundaries on the plausible range of benefits from reduced exposures. Even broad limits can be helpful in choosing strategies; these limits do not indicate the "optimal" solution, but they help avoid major mistakes.

This chapter focuses first on the health effects associated with benzene exposure—in particular, leukemia. Several dose-response models are discussed, though, for lack of a widely accepted alternative, primary reliance is on the linear model estimated by the EPA's Carcinogen Assessment Group (a model that almost certainly overestimates the risks and hence overestimates the benefits of reducing exposure). The second half of the chapter addresses how to value reductions of risk in monetary terms, drawing on the substantial recent literature on the value of life, or, as Schelling (1968) has phrased it more accurately, the value of saving a life.

Estimating the Risk

The evidence is overwhelming that long-term exposure to benzene at levels about 100 ppm poses significant risks of aplastic anemia, leukemia, and other life-threatening disorders. The evidence does not address directly the risks posed by the low-level exposures that are relevant to the EPA's decisions, though the absence of demonstrable health effects at low exposures is no proof that such exposures are harmless.

Noncarcinogenic effects

Virtually none of the noncarcinogenic health effects associated with benzene are of any concern at concentration levels found in the general environment. The concept of thresholds (exposures below which no risk is incurred) is well accepted for most noncarcinogenic effects.[1] A common aphorism among toxicologists is "The dose alone makes a poison" (Zapp 1977). For such benzene-related illnesses as aplastic anemia and other pancytopenias, we have every reason to believe that current environmental exposures are well below any thresholds.

The situation is slightly less comforting with regard to chromosomal aberrations, for which it is not clear that thresholds exist. As discussed in chapter 6, however, the evidence that benzene causes chromosomal damage is nonspecific; benzene has not been connected with any particular types of birth defects. Evidence of chromosomal aberrations does help to corroborate other evidence that benzene is a leukemogen, but that suggests that by estimating the dose-response function for leukemia we can capture virtually all of the health effects associated with low-level exposures to benzene.

Low doses and carcinogens

All of the evidence on benzene's leukemogenicity has been derived from populations exposed at levels several orders of magnitude higher than those found in the environment. No data show an increased incidence of leukemia in individuals exposed to benzene concentrations that even remotely approach the one-part-per-billion range. Some argue that the absence of such data indicates a threshold, that exposures of even tens of parts per million pose no risk of leukemia. Testifying before the EPA's scientific advisory board, James Jandl, then chairman of Hematology at Harvard Medical School, argued that benzene-induced leukemia is always preceded by aplastic anemia and that it occurs only at exposure levels exceeding 100 ppm. (See Jandl 1977.)

More generally, those who argue for thresholds emphasize the existence of protective biological mechanisms that prevent cancer development unless the dose is high enough to overwhelm these defenses. Advocates of this concept point to a variety of vitamins (e.g., D_2), metals (e.g., nickel), and minerals (e.g., calcium) that are biologically essential at low doses yet are known to be carcinogens at high doses. They conclude that thresholds must exist for many, if not all, chemical carcinogens (Maugh 1978).

Others dispute the existence of thresholds for carcinogens, or suggest that even if they do exist they cannot be identified by current experimental and epidemiological techniques (National Academy of Sciences 1977). As noted earlier, the EPA and other regulatory agencies act on the assumption that any exposure to a carcinogen poses some risk. The Delaney clause, for example, requires that the FDA ban any food additive shown to be a carcinogen in any mammalian species at any dose.

To the layman, the inability of scientists to resolve the threshold debate may seem strange. If we test animals at low doses or study human populations exposed at low levels and do not observe any increased incidence of cancer, can we not conclude that low concentrations are safe? The answer, unfortunately, is no. Consider an experiment exposing 500 laboratory animals to a particular carcinogen at a specific dose, with additional animals used as unexposed controls (a rather large experiment by current standards). Suppose no cancer is observed and, to keep matters simple, the test species normally has a zero incidence of cancer. It might be tempting to conclude that we have found a threshold, but such a conclusion would be on very shaky ground. Even if the risk were as high as 1.4×10^{-3}, there would be a 50 percent chance that no tumors would be observed in 500 animals. To ensure a 50 percent chance of detecting a risk of 10^{-6}, almost 700,000 animals would be needed.[2] Such a huge experiment would be prohibitively expensive, if not impossible to run. Moreover, if the test species had some natural background incidence of cancer, as is normally the case, the experiment would have to be larger still (Schneiderman et al. 1975). Epidemiological studies also are incapable of detecting very small risks, except where a chemical causes an otherwise rare form of cancer, as with vinyl chloride and angiosarcoma. Unless the risk is highly elevated above the baseline, it cannot be detected with any assurance (Calkins et al. 1979). Moreover, given the poor exposure date in typical studies, it is difficult to use the results to define a threshold.

Extrapolative models

Since low-level risks cannot be measured empirically, several mathematical models have been developed to extrapolate from high-dose, high-risk data to predict risks at low doses. Each model has its respected scientific champions, yet they yield radically different predicted health effects at low doses. Unfortunately, current theory fails to select any single model, nor can one be selected empirically. Moreover, most models predict only the incidence of cancer, not the years of life lost. Models that predict the time from exposure to the onset of cancer are at an even more primitive stage.

The "one-hit" model assumes, at least metaphorically, that cancer may be induced by even a single hit of a susceptible cell by a carcinogen.[3] Thus, the lifetime risk of cancer at dose D, $P(D)$, is the probability of one or more hits:

$$P(D) = 1 - \exp(-\gamma D) \tag{7.1}$$

where γ is a constant to be estimated empirically. At low doses, this function is approximated very closely by the linear expression

$$P(D) = \gamma D \tag{7.2}$$

where γ has the same value in both equations. Thus, at low doses, if the dose is reduced, the risk will fall in proportion. This model is the most conservative of those commonly proposed, giving an upper-bound estimate of risk levels at low doses.

The most extensive application of the one-hit or linear model has been to low-level ionizing radiation, where it is widely (though not universally) accepted. Some scientists argue, however, that radiation and chemical carcinogenesis are fundamentally different—that radiation acts directly on cells, but chemicals are mediated by metabolic and other processes that may make the risk less than proportional to dose at low concentrations (Maugh 1978). In contrast, Crump et al. (1976) argue that, even if the curve for any single chemical in isolation is nonlinear, if the effects of different carcinogens are additive and the burden imposed by any single chemical is small relative to the total, then the change in risk will be proportional to the change in dose for any given chemical. (In the language of economists, all of the exposure to any single chemical is at the margin.)

The other models yield curves that are convex at low doses. If the dose is reduced, the risk falls more than proportionately. The log-probit model is the most widely used such model. It has little theoretical rationale; it was developed for the safety testing of drugs and noncarcinogenic chemicals, where the data appear to be consistent with the assumption that individual tolerances or thresholds are distributed lognormally across the population. Thus the risk is given by

$$P(D) = \Phi(\alpha + \beta \log D) \tag{7.3}$$

where $\Phi(\bullet)$ is the cumulative normal distribution and α and β are constants estimated empirically. This function is convex at low doses. The larger β, the more rapidly the predicted risk level declines as the dose decreases. An important special case of the log-probit model, called the Mantel-Bryan approach, sets $\beta = 1$, so that only α must be estimated; each tenfold reduction in dose causes a shift of one standard deviation in the normal distribution. (See Mantel and Bryan 1961.)

The logit model yields an S-shaped curve similar to that of the log-probit model, but with a "fatter tail" at low doses. Predicted lifetime risk under the logit model is

$$P(D) = [1 - \exp(-\alpha - \beta \log D)]^{-1} \tag{7.4}$$

where α and β are estimated constants, and increases in β cause a more rapid falloff in risk as the dose is reduced. When estimated from the same data and extrapolated to low doses, the logit model predicts higher risks than the log-probit model but lower risks than the linear model.

The multihit model is a generalization of the one-hit model; for a k-hit model, the risk is equal to the probability that k or more hits will occur:

$$P(D) = \int_0^{\gamma D} [x^{k-1} e^{-x} / \Gamma(x)] dx \tag{7.5}$$

where $\Gamma(\bullet)$ is the gamma function, $\Gamma(k) = (k-1)!$ for k a positive integer and $\int_0^\infty x^{k-1} e^{-x} dx$ otherwise, and γ is a constant to be estimated. (If $k = 1$, equation 7.5 reduces to equation 7.1.) At small doses, the k-hit model is very closely approximated by a power function:

$$P(D) = ax^k. \tag{7.6}$$

Thus, for $k > 1$, the k-hit model is convex at low doses.

Comparing the models

All of the models yield similar curves over the range of risk levels that can be detected in laboratory and epidemiological studies.[4] Figure 7.1 plots five such dose-response curves, where the parameters have been set to yield a risk of 0.5 at an arbitrarily scaled dose of 0.5. The one-hit, logit with $\beta = 3$, and log-probit with $\beta = 2$ curves are virtually identical. The two-hit and Mantel-Bryan (log-probit with $\beta = 1$) curves are more easily distinguished from the others, but even with them the data would have to be fairly extensive and of high quality in order to indicate the appropriate model empirically.

The inability to choose among the models on experimental evidence would not be troublesome if they yielded similar results at low doses, but they do not. Suppose we know from a very careful epidemiological study that continuous exposure to a certain chemical at 1 ppm raises the lifetime risk of some form of cancer by one case per hundred people; at $D = 1$ ppm, $P(D) = 0.01$. Suppose further that environmental exposures average 1 ppb (10^{-3} ppm), a thousand times lower. To predict the risk at that exposure, we fit each of the models to our data and extrapolate, with the results in figure 7.2.

The models yield wildly different risk predictions at low doses. At 1 ppb, the one-hit model predicts a risk of 1.0×10^{-5}. The others are all much lower. Recall how similar the one-hit, logit with $\beta = 3$, and log-probit with $\beta = 2$ models looked in figure 7.1. Yet at 1 ppb, the logit model with $\beta = 3$ predicts only 1.2×10^{-6}, about 10 times lower than the one-hit model. The log-probit with $\beta = 2$ estimate is even lower—approximately 4×10^{-17}, or more than 100 billion times less than the one-hit estimate. Extrapolated far below the observed data, the one-hit model yields far higher risk estimates than any of the other models.

The CAG's dose-response function
It is impossible to derive a dose-response curve for benzene that can be defended rigorously. Any estimated curve would be open to a variety of criticisms, many of them valid. Unfortunately, the uncertainties are unlikely to be resolved within the foreseeable future; a "wait-and-see" strategy would buy little new information on low-dose risks from benzene.

In deriving its dose-response curve, the EPA Carcinogen Assessment Group adopted the conservative one-hit model.[5] It made the additional assumption that lifetime risk is proportional to cumulative exposure: 70 years of exposure at 1 ppb poses the same risk as 7 years at 10 ppb, 0.07 years at 1 ppm (1,000 ppb), or 0.007 years (2.6 days) at 10 ppm (the current occupational limit). This assumption allowed the CAG to extrapolate from workers exposed to high but intermittent concentra-

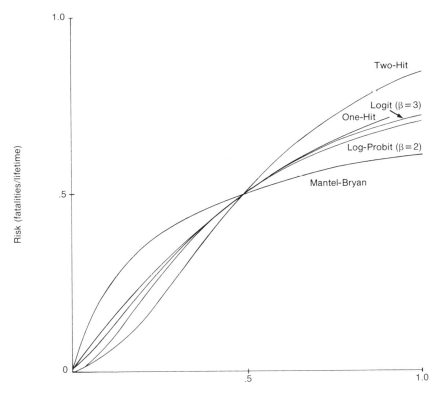

Figure 7.1   Comparison of dose-response models in the observable range.

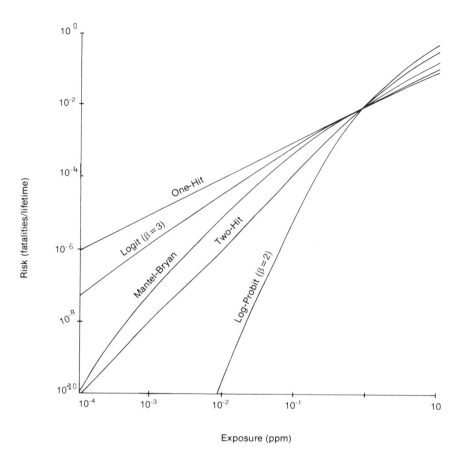

Figure 7.2 Low-dose extrapolation.

Table 7.1   Parameters and slope estimates from three epidemiological studies.

Study	Type of leukemia	P_0	R	Δx	B
Infante	All	0.006732	7.20	2.81	0.014854
Askoy	Nonlymphatic	0.004517	19.92	4.22	0.020252
Ott	Myelogenous	0.002884	3.75	0.17	0.046380

Source: Albert et al. 1979, tables 2 and 3.

tions to the general population, which is exposed to much lower levels of benzene over longer periods of time. (With any of the nonlinear models such extrapolation would be very difficult, as the distribution of exposures matters, not just their mean.)

The CAG's model takes the form

$$P = A + Bx \tag{7.7}$$

where P is the lifetime risk of leukemia, A is the risk in the absence of benzene, and x is the exposure to benzene averaged over a lifetime. The excess risk due to benzene exposure, Bx, is proportional to the average exposure factor. The problem is to estimate B.

Equation 7.7 can be put in a form that facilitates the estimation of B from studies of workers exposed to benzene:

$$B = P_0(R-1)/\Delta x \tag{7.8}$$

where P_0 is the baseline expected risk in the population, R is the ratio of observed to expected cases of leukemia, and Δx is exposure to benzene in the workplace measured in ppm. The CAG estimated R and Δx for each of three epidemiological studies (Infante et al. 1977, Aksoy et al. 1974, and Ott et al. 1977). The estimate of P_0 varied from study to study depending on the particular type of leukemia for which R was calculated. Table 7.1 summarizes the parameter values used by the CAG for each of the three studies and the resulting estimates of B, which range from 0.015 to 0.046. The CAG's final estimate is simply the geometric mean of the three:[6]

$$B = [(0.014854)(0.020252)(0.046380)]^{1/3} = 0.024074. \tag{7.9}$$

The estimate of $B = 0.024$ represents deaths per lifetime per ppm exposure. It implies that if 1,000 people were exposed to 1 ppm benzene over their lifetimes, about 24 of them would die prematurely from leukemia. It is more convenient, however, to express the risk in deaths per ppb-person-year, dividing B by 70.96 (the expected lifetime in the life tables used by the CAG) and by 1,000 (the number of ppb per ppm). That yields 0.339×10^{-6} deaths per ppb-person-year (Albert et al. 1979, p. 21).

The CAG also attempted to quantify the uncertainty in its estimates. Under certain very restrictive assumptions, the 95 percent confidence interval around the estimate of 0.339×10^{-6} ranges from 0.129×10^{-6} to 0.887×10^{-6}. Various EPA documents (e.g., U.S. Environmental Protection Agency 1980) cite these limits as "low" and "high" estimates, but they have little meaning because they ignore the major uncertainty: the appropriate extrapolative model. Other models would yield estimates orders of magnitude lower.

Even if the linear model is the correct one, the CAG estimate is open to question. Two EPA analysts, Luken and Miller (1979), have criticized the CAG's interpretation of the epidemiological data—in particular, the estimates of exposure levels and the inclusion of deaths of workers who were not in the original cohorts studied or who had been exposed to other suspected carcinogens besides benzene. They argue that 0.081×10^{-6} deaths per ppb-person-year (one-fourth of the CAG figure) is a more reasonable estimate.

Lamm (1980), an occupational physician and consultant to the American Petroleum Institute, also has criticized the CAG's assumptions. He offers many of the same criticisms as Luken and Miller, and also argues that the CAG overestimated the relative risk factor (R) and underestimated the exposure level in one of the epidemiological studies. Using the linear model, he estimates the risk to be 0.0327×10^{-6} deaths per ppb-person-year, more than 10 times lower than the CAG.

Choice of risk factor
With a wide range of alternatives, the ideal decision-analytic approach would be to assess the probabilities that the various models and epidemiological estimates are "correct" and then use those probabilities to compute an expected risk function. Unfortunately, this course is not feasible here. We have estimates for only the linear model, and even if we had others we have no way of assessing the likelihood that the various models are correct. It might be useful to convene a panel of experts to undertake that task (perhaps under the auspices of the National Academy of Sciences), but that is well beyond the scope of this study.[7] We can, however, discuss the general shape of such an expected risk function. The nonlinear models predict near-zero risk at low doses. If we assign any significant probability to the linear model's

being correct, the expected risk function also will be linear, albeit with a smaller coefficient than the linear model's.[8] Thus, it seems reasonable to proceed with the assumption that expected damages will be proportional to exposure.

We still have to determine what coefficient of proportionality to use. Often it is useful to employ a range of estimates and to perform extensive sensitivity analyses. When there are uncertainties about several aspects, however, the presentation of calculations becomes immensely complicated, overwhelming the reader with an impenetrable mass of alternative estimates. Thus, the analyses of regulatory options in the next two chapters will employ the CAG risk estimate of 0.339×10^{-6} deaths per ppb-person-year, without sensitivity analyses. Although conservative assumptions are generally an inappropriate basis for policy, two factors recommend use of the CAG estimate in this context: Our main interest is the comparative advantages of alternative regulatory strategies, not the detailed costs and benefits of regulating benzene *per se*, and even with the CAG estimate the standards under consideration do not pass any reasonable benefit-cost test.

The Value of Saving a Life

To compare the benefits with the costs of controlling benzene exposure, we must express the two in commensurate terms. As suggested already, the dominant benefit is likely to be a reduction in premature deaths. The nonleukemic risks of low-level exposures appear minuscule, and leukemia in adults is almost invariably and rapidly fatal. Thus, the central problem is one of valuing the saving of a life.

For many individuals, the very idea of placing a dollar value on human life is offensive: "Life is priceless." Yet, while life may indeed be priceless, the amount that we are willing to spend to avert fatal risks, as a society and as individuals, is finite. We could, after all, probably save virtually all of the 50,000 or so lives lost in motor-vehicle accidents each year by lowering the speed limit to 20 miles per hour. Similarly, most deaths from fire could be averted if we banned wood-frame construction and required that all buildings and furnishings be made of nonflammable materials. That we do not take these actions (and many others) suggests that there are limits on what we are willing to spend to save a life.

A large literature, both conceptual and empirical, has grown up around the question of valuing life-saving and health-promoting activities. The discussion that follows highlights the conceptual issues and presents empirical results from studies or risky occupations and other sources.

Willingness to pay

A consensus appears to have emerged, at least among economists, that the value attached to reducing risk should be based on the criterion of willingness to pay. The principle is simple: The value of some benefit to an individual is the amount he would be willing to spend to secure it. In its most basic form it is almost tautological, but it offers some useful insights and general guidance even though it does not provide a single, unambiguous dollar estimate.

The assessment of willingness to pay focuses on the valuations of those individuals affected most directly, the people whose lives may be saved by the program under consideration. The relevant question typically is not how much an individual would pay to avoid certain death, but how much he would pay to reduce the probability of premature death by some small amount. In the first instance, the individual, if he had to pay the cost himself, would be restricted to an amount no greater than his discounted earnings. In the second case, however, income is unlikely to be a binding constraint, though no doubt it will be an important factor. (As Zeckhauser 1975 suggests, some individuals might be willing to give up 10 percent of their future income in order to eliminate a 1 percent chance of immediate death.)

Some readers may be uncomfortable with formulating the question in terms of how much an individual would pay to reduce an environmental risk that is imposed on him. An alternative is to ask how much compensation the individual would have to receive to be willing to accept the risk. Here again, the distinction between certain death and a low probability of death is crucial. If we were to ask an individual how much compensation he would require to accept certain, immediate death, most likely the answer would be that no amount of money would be sufficient. If we were to ask, however, how much he would require to accept some small probability of death, most likely the answer would be a finite amount. In fact, economic theory suggests that for very small risks, the amount of compensation required will be almost identical to the amount that a person would be willing to spend to eliminate an existing risk of equal magnitude.[9]

The rationale for willingness-to-pay measures is the concept of a potential Pareto improvement—a change that makes some people better off without worsening anyone's position. Consider a hypothetical situation where exposure to some airborne chemical imposes an annual risk of death of 10^{-6}. Ten million people are affected, and each would be willing to spend up to $1 per year to eliminate that risk. At a cost of $5 million per year, industrial users of the chemical can switch to a safe alternative. Under the willingness-to-pay criterion, the government should require them to do so, as each of the 10 million people could chip in 50 cents to compensate industrial users yet still be better off.

Now suppose the cost of the substitute is $15 million instead of $5 million. The government proposes a ban. Industry discovers that each of the 10 million individuals is indifferent between a payment of $1 per year and imposition of the ban. Under the willingness-to-pay criterion, the ban should not be imposed; if the chemical remains in use, industry could pay each individual $1, making him as content as he would have been had the ban been imposed yet saving itself $5 million per year. The problem with this argument is that the compensation might never be paid. The individuals at risk might argue that, although they would gladly accept compensation for forgoing the ban, if such compensation would not be made the ban should be imposed. This criticism of willingness to pay is not unique to life saving; it arises in many contexts.

The distribution-based argument against the willingness-to-pay criterion is least compelling when the potential costs and benefits are widely distributed across the population, as they are with airborne benzene. The ubiquity of benzene in gasoline ensures that exposure is widespread, so that the associated risks are not borne disproportionately by any single group (though obviously variations do exist). The ultimate financial burden of controlling benzene also would be spread widely. Although the costs would be borne initially by a small number of large chemical and petroleum firms, most would be passed on to consumers in higher prices. In view of the wide range of products produced with benzene and the near-universal use of gasoline, virtually everyone would end up sharing at least a portion of the costs. Thus, the case for willingness to pay appears to be quite strong.

Empirical estimates
The major obstacle to the full implementation of the willingness-to-pay criterion lies in the difficulty of estimating the appropriate dollar amounts. Various techniques, some quite ingenious, have been proposed and tried. Although none has proved fully satisfactory, together they provide useful inputs for decision makers.

As Schelling (1968) suggests, one approach is to ask people. Acton (1973) asked a small sample of individuals how much they would pay to have a mobile cardiac unit that would reduce the fatality rate among heart-attack victims. When the respondents were told to assume that the unit would lower the risk of death by some specified amount (10^{-3} or 2×10^{-3}), their answers suggested implicit values per life saved in the range $28,000–$43,000. The questions were hypothetical, however, and most individuals have difficulty dealing with small probabilities (Tversky and Kahneman 1974), so little faith can be put in these estimates.

Other attempts have been indirect, drawing inferences from actual behavior in labor markets and elsewhere. An early, well-known study by Thaler and Rosen (1976) looked at the wage premiums in high-risk occupations. Their estimates suggest that workers in their sample demanded about $200 extra pay per year for each 0.001 increase in the annual risk of death, implying a "value per life saved" of about $200,000 in 1967 dollars. As Zeckhauser (1975) and Viscusi (1978) point out, however, workers in high-risk occupations are likely to include a disproportionate number of individuals with relatively low valuations, so the results may not be applicable to the population as a whole.

Smith (1976), Viscusi (1978), and Dillingham (1979) estimated wage premiums for hazardous work using broader, more representative samples of workers. Smith, using industry-level data, found wage premiums of about $2.6 million per life in his 1967 sample and $1.5 million per life in his 1973 sample. Although Smith suggests that for a variety of reasons the second (lower) estimate is probably more reliable, it is still higher than that of Thaler and Rosen by almost an order of magnitude. Viscusi estimates similarly substantial wage premiums, in the range of $1 million to $1.5 million per life in 1969–70 dollars. In contrast, Dillingham estimates much lower wage premiums using similar data. In 1969 dollars, he estimates a wage premium for blue-collar workers of $368,000 per life, which falls to $168,000 when adjusted for the risk of injury (as cited in Bailey 1980, p. 44).

Blomquist (1977) estimated willingness to pay as revealed by individual choices with regard to an activity explicitly related to safety: the use of seat belts. Blomquist's procedure is somewhat complex and indirect, as the cost of using seat belts that are already in the car is in time rather than dollars. For the average driver, he estimates an implicit value between $142,000 and $488,000, with an "intermediate" estimate of $257,000 in 1975 dollars (as cited in Bailey 1980, p. 40).

Bailey (1980) uses some of these estimates to derive a range of "values of life" that he considers appropriate for use in evaluating risk-reduction programs. His final "intermediate" estimate is $360,000 per life in 1978 dollars, with "low" and "high" estimates of $170,000 and $715,000 (Bailey 1980, p. 46). Those estimates are based on the studies by Thaler and Rosen, Blomquist, and Dillingham, although Bailey adjusted the estimates from each study to achieve greater consistency in underlying assumptions and to put all of the estimates in 1978 dollars. Bailey's results, however, do not incorporate the estimates of Smith or Viscusi. If those were included, the range would be greatly expanded on the upward side, to several million dollars per life. Bailey argues against the use of those high estimates, partly because he believes that the aggregate data used by Smith and Viscusi tend to make their estimates less accurate. Perhaps more compelling is Bailey's argument that their high estimates yield implausible implications for individual behavior.

Bailey offers the following example. Suppose a family of four with an income of $18,500 were offered an opportunity to reduce the annual risk of death for each of its members by 0.0005 (which is about the decline in U.S. death rates from 1970 to 1975). If the value of saving a life is V, then the family should be willing to pay up to $(4 \times 0.0005)V = 0.002V$ dollars per year. Using Bailey's intermediate estimate of $V = \$360,000$, the family would pay up to $720—about 4 percent of its income. If we inflate Smith's high estimate to 1978 dollars, however, then $V = \$5$ million, and the family should be willing to spend up to $10,000 per year—over half its income (Bailey 1980, pp. 45–46). It is hard to disagree with Bailey that this result is highly implausible. Less extreme figures of $1 million or $2 million, however, which also are well beyond Bailey's range, cannot be rejected so confidently.

Length and quality of life
A major difficulty with these estimates is that they fail to reflect the length and quality of the lives saved. Death is inevitable; programs prolong lives rather than save them. Consider two programs that will save the same number of lives. One will save healthy 30-year-olds and the other will save seriously ill 75-year-olds. Surely most people would be willing to spend more money on the first program than on the second, though if we define benefits purely in terms of lives saved the two yield equal benefits.

Zeckhauser and Shepard (1976) argue that we should measure "quality-adjusted life years" saved, to take account of both the number of years saved and such factors as physical disability. This measure is far from operational, but it does have some qualitative implications. The estimates for benzene predict only the number of deaths, and not their distribution by age. We know, however, that cancers disproportionately affect the elderly; the death rate for myelogenous leukemia (the type most commonly associated with benzene) is 26 times higher among people aged 70–74 than among those aged 1–5 years (Albert et al. 1979, table 1). This suggests that if benzene causes a proportional increase in leukemia rates across age groups, those saved by reducing benzene exposure will be considerably older than the average U.S. resident. Thus, we might wish to spend less money per life saved on benzene than we do on other programs, such as highway safety, that tend to have a more even impact on different age groups.

Discounting
Once we think of years saved, the question of discounting naturally arises. In benefit-cost analyses of projects that generate a stream of costs and benefits over time, the standard procedure is to discount both streams to compute the net present value. The discount rate represents the opportunity cost of the funds employed. A widely debated issue in the economic literature on life saving is whether "lives saved" or "years of life saved" should be discounted in the usual way.[10] A simple example sharpens the issue: Consider two programs that each cost $100 million today. Program A saves 100 40-year-olds from immediate death in auto accidents. Program B reduces the exposure of 20-year-olds to a carcinogen; without the program, 200 of them will die from cancer in 20 years at age 40. If we do not discount, program B is clearly preferable; it offers double the benefits ("saved" 40-year-olds). If we discount, however, program A may well be preferred. At a 5 percent real discount rate (low by most standards), $1 received 20 years from now is worth only $0.377 today. If we apply that discount factor to program B, the net present value is only $0.377 \times 200 = 75$ lives; thus, program A should be chosen.

Critics of discounting argue that it discriminates against future citizens.[11] Failing to discount, however, leads to some peculiar conclusions. Consider a third option: Under program C, the $100 million is invested, earning 5 percent. In 20 years, the money accumulated will be used to implement an expanded version of program A. The total amount available at that time will be $(1.05)^{20} \times 100 = \$265$ million. If program A can be expanded proportionally, 265 40-year-olds can then be saved from death in auto accidents. If we do not discount, program C is clearly preferable to A or B. But then why not leave the money invested for an even longer time, so that even more lives can be saved later? Carried to its logical conclusion, the no-discounting position argues for investing all available funds for life saving in the indefinite future, rather than spending them immediately on programs that yield benefits only after a long delay.

This issue is extremely important to the question of how we should value the lives saved by reducing benzene exposure. The lag times between exposure to a carcinogen and appearance of disease can be on the order of 20 years or more. Thus, reducing benzene exposure may not yield many benefits until far in the future. If we combine discounting with the idea that life years saved rather than lives saved is the appropriate benefit measure, the willingness-to-pay estimates may need to be scaled down. Suppose we have an estimate of $1 million, based on an immediate risk of death for a 40-year-old who can expect to live for 35 more years. At a 5 percent discount rate, that would imply a value of $58,164 per year of life.[12] How should that estimate be used to compute the benefit of reducing benzene exposure for the same individual when doing so would reduce the risk of leukemia 15 years from now, saving the last 20 years of life? If we ignore years of life and discounting, each life saved should be valued at $1 million. If we use that implicit value per life year and discount at 5 percent, however, each life saved should be valued at only $366,100.[13] These numbers are hypothetical. We do not have estimates of the lag time between benzene exposure and the onset of leukemia. We do, however, have reason to believe that it may be substantial, so it may be appropriate to use a lower value per life saved than for programs that immediately save the lives of younger people.

Summary

Massive uncertainties pervade estimates of the risks of benzene exposure and the present dollar value of "saving a life." Because of the nature of the uncertainties involved, alternative estimates cannot be collapsed into a single function that predicts the expected damages of environmental exposures to benzene. Despite the uncertainties, we can draw two tentative conclusions about the expected benefits of reducing benzene exposure. These conclusions provide important guidance in assessing alternative regulatory strategies.

First, the expected benefits of reducing ambient benzene concentrations probably are proportional to the reduction in total exposure. Even if we assign only a modest probability to the linear model's being correct, the expected dose-response function will be linear at low doses given the minuscule risks predicted by the nonlinear models. With benefits proportional to the reduction in exposure, the optimal degree of control depends critically on the costs of control. Cutting exposures from 2 ppb to 1 provides the same expected benefits as cutting them from 4 to 3 or from 1 to 0. There is no target level of exposure.

Second, the benefit is almost certainly no more than $1 per ppb-person-year and is probably considerably lower. The CAG's estimate probably overstates the true risk, yet it predicts only 0.339×10^{-6} fatalities per ppb-person-year. If we combine the CAG's estimate with a value per life saved as high as $3 million, the value of eliminating one ppb-person-year of exposure is $(0.339 \times 10^{-6})(\$3 \times 10^6) = \$1$. More plausible assumptions generate considerably smaller estimates. If we assume, for example, that the true risk is half that estimated by the CAG (but double that estimated by Luken and Miller and five times higher than Lamm's estimate) and that the present value of averting a future fatality due to current benzene exposure is $500,000, cutting exposure by one ppb-person-year is worth only $(0.17 \times 10^{-6})(\$5 \times 10^5) = \$0.085$. With Lamm's risk estimate and Bailey's "low" estimate of $170,000 per life saved, the benefit falls to $(3.27 \times 10^{-8}) \times (1.7 \times 10^5) = \$0.0056$ per ppb-person-year.

Chapter 8

Maleic Anhydride Plants

Maleic anhydride plants were the target of the first "benzene regulation" proposed by the Environmental Protection Agency (45 *Fed. Reg.*, 26660, 1980). The implicit logic behind this choice is clear: The emissions study (PEDCo Environmental, Inc. 1977) indicated that over half of the benzene emissions from chemical manufacturing came from fewer than ten maleic anhydride plants. Within each plant, virtually all of the emissions were released from a single point source—the product-recovery absorber—for which several well-known, highly efficient control techniques are available. The *prima facie* case for regulation was strong. This chapter examines in detail the costs and benefits of the proposed regulation and some alternatives.

The proposed standard requires that maleic anhydride plants monitor emissions continuously. This mutes one of the major practical objections to the economist's prescription of a charge levied on emissions.[1] However, several factors suggest that a classic, uniform emission charge is likely to be far from optimal and that some alternatives merit consideration.

The efficiency advantage of an emission charge over a standard is greatest when control costs vary widely across sources. Control costs are likely to vary most when the regulation covers different industries with different production technologies and control techniques. Within a given industry, control costs may vary widely if firms employ different production processes, or if controls are subject to economies of scale and sources vary widely in size. None of these conditions applies here; a single industry is involved, the production processes are fairly uniform in the plants affected, economies of scale in control are modest, and plant capacities do not vary dramatically. These facts suggest that the ability of charges to allocate control efforts differentially based on costs may be relatively unimportant.

In contrast, plants vary widely in estimated health effects per unit of emissions. One is in a large city, another is in a lightly populated rural area. Estimated exposures per kilogram emitted vary across plants by a factor of almost 50. This variation appears to be a more significant factor in the selection of cost-effective strategies than variation in control costs. If the relationship between emissions and damages varies across sources, a simple uniform charge on emissions will not be optimal. Such a charge equalizes the marginal costs of controlling *emissions* at different plants. Efficiency requires, however, that the marginal costs of controlling *damages* be equalized; it is worth more to control emissions from plants at high-damage sites. Thus we need to consider charges and standards that vary with marginal damages.

The main body of this chapter is based on data available to the EPA in the spring of 1980 and was completed before the EPA held a public hearing on the proposed standard in August 1980. Witnesses at that hearing offered important new information and alternative estimates of control costs. A postscript analyzes that new information, which generally reinforces the chapter's major conclusions.

Overview of the Maleic Anhydride Industry

Production and use

Maleic anhydride is produced in ten plants in the United States, with a total annual capacity of approximately 229 million kilograms (about 500 million pounds) (U.S. Environmental Protection Agency 1980, p. 1-3). Although capacity utilization was only 56 percent in 1977, the EPA projects that production should reach about 95 percent of current capacity by 1982 (U.S. Environmental Protection Agency 1980, p. 1-1). Slightly over a half of the maleic anhydride produced is employed in the manufacture of unsaturated polyester resins, which in turn are used in a variety of reinforced plastics. Other major uses of maleic anhydride are for agricultural chemicals (10 percent), lubricating additives (7.8 percent), and fumaric acid (6.4 percent) (U.S. Environmental Protection Agency 1980, p. 5-8). Three production methods are employed in the United States: oxidation of benzene, oxidation of *n*-butane, and recovery of maleic anhydride as a byproduct of the manufacture of phthalic anhydride. These processes account, respectively, for 82, 16, and 2 percent of total capacity. Of the ten plants, only one recovers maleic anhydride in manufacturing phthalic anhydride, and one uses the *n*-butane process exclusively. Neither of these plants emits any benzene. The eight remaining plants employ the benzene oxidation process, although in one the *n*-butane process accounts for 20 percent of capacity (U.S. Environmental Protection Agency 1980, p. 1–3).

The benzene oxidation process

In the benzene oxidation process, benzene is fed into a reactor, where it is oxidized, yielding maleic anhydride, water, and other substances in gaseous form. The stream of gas leaving the reactor is partially condensed, and an initial separation yields about 40 percent of the maleic anhydride ultimately produced. The remaining gas is routed to a product-recovery absorber, where it is converted to liquid form for additional processing. Approximately 99 percent of the benzene emitted from maleic anhydride plants is released from the product-recovery-absorber vent (U.S. Environmental Protection Agency 1980, pp. 1-6-1-11).

A plant's rate of uncontrolled emission from its recovery-absorber vent depends primarily on the conversion rate of benzene in the production process. In its analysis, the EPA assumed that all plants achieve 94.5 percent conversion, resulting in an emission rate of 0.067 kg benzene per kg maleic anhydride (U.S. Environmental Protection Agency 1980, p. 1-10). Plant-specific conversion rates, however, vary between 90 and 97 percent (U.S. Environmental Protection Agency 1980, p. 1-7), so uncontrolled emission rates actually range from 0.037 to 0.122 kg. Very minor amounts of benzene also are emitted from fugitive sources, such as leaks in pipes and valves. Benzene accounts for almost 80 percent of the total emissions of volatile organic chemicals from benzene-fed maleic anhydride plants; formaldehyde, maleic acid, and formic acid make up the remainder (U.S. Environmental Protection Agency 1980, p. 1-12).

Control techniques

Five of the eight plants currently employ some form of emission control, primarily to meet state hydrocarbon regulations. Carbon adsorption and thermal incineration, described below, are each used at two plants. One plant uses a catalytic incinerator, where a catalyst permits destruction of hydrocarbons at relatively low temperatures. The EPA estimates that the plant with a catalytic incinerator (U.S. Steel) achieves 90 percent control efficiency, while the two plants with thermal incinerators achieve 97 (Denka) and 99 percent (Koppers). The carbon adsorber systems achieve estimated efficiencies of 90 percent at one plant (Reichold, Illinois) and 97 percent at the second (Reichold, New Jersey). Three plants (Ashland, Monsanto, and Tenneco) have no controls (U.S. Environmental Protection Agency 1980, p. 1-13).

Table 8.1 summarizes this information for the eight plants . The column labeled "uncontrolled" is based on the EPA's emission factor of 0.067 kg benzene per kg maleic anhydride and assumes full-capacity operation. The column labeled "Current Control" reflects the impact of existing controls, if any; for example, the Denka plant, which achieves an estimated 97 percent control, has emissions that are 3 percent of the "uncontrolled" level.

The EPA has focused its analysis on the product-recovery-absorber vent. Control techniques fall into three major categories: (1) modifications of the benzene-fed production process (for example, employing a different catalyst that yields less unreacted benzene and hence a lower level of emissions, but also less maleic anhydride, (2) substitution of an alternative feedstock (almost certainly *n*-butane), thus eliminating benzene, and (3) installation of add-on control devices that remove most of the benzene from the exhaust stream. Unfortunately, the EPA has developed cost estimates only for add-on controls, and only for a very restricted control range. Early in its analysis, the agency appears to have limited its investigations to those options that achieve high levels of control. Modifications of the benzene oxidation process without add-on controls appear to be incapable of reducing benzene emissions to comparable levels. They can, however, achieve more moderate reductions, possibly at low cost. For example, a plant that increased its benzene conversion rate from 94 to 97 percent would in effect cut its emissions in half. Similarly, commercial-scale catalytic incinerators have achieved removal efficiencies of "only" 85–90 percent (U.S. Environmental Protection Agency 1980, p. 2-4). The *n*-butane process, which eliminates benzene emissions entirely, is new enough that firm cost estimates are unavailable (U.S. Environmental Protection Agency 1980, p. 1-11); much of the required information is proprietary. The EPA has not analyzed the costs of conversion to *n*-butane, though its proposed zero benzene-emissions standard for new plants would in effect force them to adopt that process.

The EPA has limited its analysis of add-on controls to thermal incineration and carbon adsorption and, for each system, to 97 and 99 percent removal efficiency. These nominal levels of control overstate the net control achieved, since the EPA estimates that controlled plants will achieve a conversion rate of 90 percent rather than 94.5 percent. Since the amount of benzene emitted increases as the conversion rate falls, the "97 percent" control hypothesized by the EPA in fact achieves

Table 8.1 Plant locations, controls, and current emissions.

Plant	Location	Capacity (million kg/yr)	Current control method	Current control level (%)	Estimated benzene emissions[a] (million kg/yr)	
					Uncontrolled	Current
Ashland	Neal, WV	27.2	None	0	1,821	1,821
Denka	Houston, TX	22.7	Thermal incinerator	97	1,520	45.6
Koppers	Bridgeville, PA	15.4	Thermal incinerator	99	1,031	10.3
Monsanto	St. Louis, MO	38.1[b]	None	0	2,551	2,551
Reichold	Morris, IL	20.0	Carbon adsorption	90	1,339	133.9
Reichold	Elizabeth, NJ	13.6	Carbon adsorption	97	911	27.3
Tenneco	Fords, NJ	11.8	None	0	790	790
U.S. Steel	Neville Island, PA	38.5	Catalytic incinerator	90	2,578	257.8
Total		187.3			12,541	5,637

Source: U.S. Environmental Protection Agency 1980 (tables 1-2 and 1-5) and author's calculations.
a. All estimates assume full-capacity operation and uncontrolled emission rate of 0.067 kg benzene per kg maleic anhydride.
b. Capacity shown for Monsanto reflects only the portion based on benzene feedstock. Total capacity is 48×10^6 kg.

a net reduction in benzene emissions of 94.55 percent. Similarly, the "99 percent" control option achieves a net reduction of 98.18 percent.[2]

Thermal incineration consists of routing the exhaust gas from the product-recovery absorber to an incinerator. Removal efficiency varies with the temperature and the residence time of the gases in the incinerator. Capital costs are little affected by the control level, but fuel requirements rise rapidly as control efficiency rises. Heat from the incinerator may be recovered for use in the plant (U.S. Environmental Protection Agency 1980, pp. 2-10–2-14).

Unlike incineration, carbon adsorption recovers for reuse most of the benzene removed from the waste-gas stream. The gas stream from the product-recovery absorber is first "scrubbed" to remove certain chemicals, and then heated before being passed through carbon beds. After several hours, the beds become saturated and must be regenerated with steam while a second set of beds is put on line. The steam is then condensed and decanted, yielding benzene and liquid waste. The net cost of the system is sensitive to the price of benzene; the higher this price, the larger the "credit" for recovered benzene (U.S. Environmental Protection Agency 1980, pp. 2-5–2-10).

Plant-Specific Costs and Benefits of Control

We can analyze three different control options for each plant using the available data: no new controls, 97 percent control, and 99 percent control. The EPA's proposed standard would limit emissions from existing plants to 0.3 kg benzene per 100 kg benzene feedstock, which corresponds to 97 percent control and a 90 percent conversion rate. New plants would not be permitted to emit any benzene. The analysis here is restricted to existing plants because at current benzene prices new plants are expected to adopt n-butane regardless of regulation (U.S. Environmental Protection Agency 1980, p. 5-4). The costs of control can be estimated in dollars, albeit with considerable uncertainty. However, the benefits of control in dollars are problematical, as discussed in chapter 7, so the estimates in this section will be stated in "lives saved." In the analysis of regulatory options, the implications of alternative dollar values will be tested.

Costs of controlling emissions
Ideally, we would like plant-specific cost estimates for a wide range of control techniques and levels. In practice, although we are dealing with

only a small number of plants, detailed plant-by-plant analyses cannot be performed. Such analyses would be very costly and would require proprietary information.

The EPA and one of its contractors have developed a hypothetical model plant that is representative of the actual plants in technology and size. The model plant's capacity of 22.7 million kg per year places it between the median and the mean of existing plants, which range from 11.8 to 38.5 million kg.[3] Once the design of the model plant was established, engineering studies of the cost and effectiveness of controls were performed. The two alternatives studied—incineration and carbon adsorption—are widely used in similar applications, so performance data are available. For a particular control design, initial costs were estimated by adding up the prices of the individual components, including labor for installation. These capital costs then were amortized over the life of the equipment (assumed to be 10 years) at an interest rate of 10 percent, and combined with estimated operating costs to yield an estimate of total annualized costs.

Crude cost estimates can be derived for individual plants using the figures for the model plant. In making such estimates, the EPA's contractor assumed that all plants were identical except for production capacity (Lawson 1978). The contractor then divided the model plant's costs into three categories: fixed costs, which do not vary with capacity or production; operating expenses, which vary in direct proportion to production; and capital costs, which vary with capacity but not with production (U.S. Environmental Protection Agency 1980; pp. 5-16–5-18).

The EPA's most recent estimates suggest that carbon adsorption has a lower net cost than incineration for all of the plants. Earlier estimates showed the reverse, but a sharp rise in the price of benzene increased the value of the benzene recovered by adsorption.[4] (This rapid rise in the price of benzene should also increase the attractiveness of conversion to *n*-butane.) In light of the apparent cost advantage of carbon adsorption, only it will be analyzed here.

Table 8.2 presents the EPA's estimates. All of the estimates include a 30–40 percent "retrofit penalty," reflecting the higher costs of adding controls to existing plants (U.S. Environmental Protection Agency 1980, p. 5-16). The total costs for all eight plants are $2.0 million and $2.7 million, respectively, at the different levels of control. These costs assume full-capacity operation (8,000 hours per year). Net annual costs are actually slightly higher if plants operate at less than full capacity; the

Table 8.2 Estimated plant-specific control costs.

| | Annual control cost[a] | | |
| | 97% | 99% | |
	Total	Total	Marginal
Ashland	410	425	15
Denka	0[b]	369	369
Koppers	0[b]	0[c]	0[c]
Monsanto	542	560	18
Reichold (IL)	320	333	13
Reichold (NJ)	0[b]	245	245
Tenneco	213	224	11
U.S. Steel	547	565	18
Total	2,032	2,721	689

Source: U.S. Environmental Protection Agency 1980, tables 5-5a and 5-6a.
a. All cost estimates are based on "full-capacity" operation (8,000 hours/year) and are in thousands of dollars.
b. Plant currently meets or exceeds 97 percent standard.
c. Plant currently meets or exceeds 99 percent standard.

small savings in operating costs are more than offset by the reduction in the credit for benzene recovered.[5] Since three of the plants already meet or exceed 97 percent control, they would incur no additional costs at that level. The EPA believes that the Koppers plant would not need any additional controls to meet a 99 percent standard, but the issue is moot, as that plant has been taken out of operation (U.S. Environmental Protection Agency 1980, p. 5-4). The EPA has assumed, however, that if a 99 percent standard were imposed Denka and Reichold (New Jersey) would have to install entirely new control equipment, and that under either option the two plants that are currently at 90 percent—Reichold (Illinois) and U.S. Steel—would require all-new control equipment. These assumptions seem pessimistic; some of the control equipment now in use might be adapted to meet the tighter controls. The costs ought to be reduced by some savings in operating costs that would be obtained when the existing systems were shut down.

Benefits of control

To predict the benefits, we must estimate the reduction in emissions, the effect of reduced emission levels on exposure, and the health gains from reduced exposures. The EPA has assembled the necessary data; most are based on the model plant, though some are plant-specific.

Table 8.3 Reductions in emissions.

	Estimated reductions in benzene emissions from absorber[a] (kg/yr)		
	From current to 97%[b]	From current to 99%[b]	From 97% to 99%[b]
Ashland	1,722	1,788	66.2
Denka	0	18	18
Koppers	0	0	0
Monsanto	2,412	2,505	92.8
Reichold (IL)	60.9	109.6	48.7
Reichold (NJ)	0	10.8	10.8
Tenneco	747	776	28.7
U.S. Steel	117	211	93.7
Total	5,059	5,417	359

a. Reductions reflect net efficiencies of 94.55% for 97% and 98.18% for 99%.
b. Change in control level, in thousands of dollars.

The estimated reductions in emissions in table 8.3 are measured relative to "current control" emissions. Thus, no reductions are shown for Denka, Koppers, and Reichold (New Jersey) at 97 percent or for Koppers at 99 percent. The third column shows the marginal reductions in emissions achieved by tightening control from 97 to 99 percent.

Translating reduced emissions into reduced exposure requires information on how emissions are dispersed in the atmosphere and on population patterns around plants. The EPA has sponsored several studies to make crude exposure estimates for each plant. One used a dispersion model and meteorological data from Pittsburgh to estimate annual concentrations of benzene at various distances from the model plant. These estimates were then scaled up or down in proportion to each plant's estimated emissions, but no adjustments were made for variations in the meteorological conditions around the individual plants (U.S. Environmental Protection Agency 1980, appendix E). Another study estimated the numbers of people residing within various distances of each plant. The contractor drew eight circles of different radii around each plant; the largest was 20 km, the greatest distance for which dispersion-modeling results were available. Using population density from the 1970 census, the contractor then estimated the population residing in each ring.

By combining population estimates with the dispersion-modeling results, we can estimate ppb-person-years of exposure per kg benzene

Table 8.4   Estimated reductions in exposure.

	Exposure factor (ppb-years/ 10^3 kg)	Estimated annual reduction in exposure (10^3 ppb-years)		
		From current to 97%	From current to 99%	From 97% to 99%
Ashland	16.8	28.9	30.0	1.1
Denka	248	0.0	4.5	4.5
Koppers	162	0.0	0.0	0.0
Monsanto	391	943.0	979.3	36.3
Reichold (IL)	8.13	0.51	0.91	0.40
Reichold (NJ)	384	0.0	4.1	4.1
Tenneco	195	145.6	151.2	5.6
U.S. Steel	179	21.0	37.8	16.8
Total		1,139.1	1,207.8	68.8

emitted. The results are given in table 8.4.[6] These "exposure factors"(per 1,000 kg) range from 8.13 for Reichold (Illinois) to 391 for Monsanto (St. Louis). That is, a kilogram emitted from Monsanto or from Reichold (New Jersey) results in almost 50 times as much exposure as one emitted from Reichold (Illinois). The other columns in table 8.4 report the estimated reductions in exposure for each plant.

The final step is to translate the reductions in exposure into estimated reductions in fatalities. As mentioned above, the EPA's Carcinogen Assessment Group estimated the risk of leukemia from benzene exposure to be 0.339×10^{-6} per ppb-person-year. Applying that factor to the estimates in table 8.4 yields the plant-specific estimated reduction in mortality shown in table 8.5. (The results assume full-capacity operation; reduced operation reduces benefits.) Monsanto alone accounts for over 80 percent of the benefits under either control option.

Uncertainties in the estimates
Neither the cost estimates nor the benefit estimates derived above are precise. Many factors contribute to the uncertainty, particularly at the level of the individual plant. Unfortunately, the nature of the uncertainty makes any attempt to define the ranges of uncertainty quantitatively almost entirely speculative or even misleading. As discussed in chapter 7, for example, the CAG's confidence limits on its estimated-risk coefficient are far too narrow and may convey a false sense of accuracy. Although it is impossible to quantify the uncertainties, it is possible to

Table 8.5 Estimated reductions in fatalities.

	Estimated annual "lives saved"		
	From current to 97%	From current to 99%	From 97% to 99%
Ashland	0.00981	0.01019	0.00038
Denka	0.0	0.00151	0.00151
Koppers	0.0	0.0	0.0
Monsanto	0.31992	0.33224	0.01232
Reichold (IL)	0.00017	0.00031	0.00014
Reichold (NJ)	0.0	0.00140	0.00140
Tenneco	0.04941	0.05131	0.00190
U.S. Steel	0.00712	0.01281	0.00569
Total	0.38643	0.40977	0.02334

identify their major sources and, in some cases, the likely direction of the bias.

The plant-specific benefits are highly uncertain. The estimates assume uniform uncontrolled emission rates and full-capacity operation, when in fact the rates vary and most plants operate well below capacity. The exposure estimates are subject to these uncertainties, plus those associated with the exposure modeling.[7] The latter include the imprecision of the dispersion model, the use of the same meteorological conditions for all plants, the assumption that time-weighted population density is the same as residential density, and the fact that concentrations indoors (where people spend most of their time) may differ from those outdoors. The greatest uncertainty arises at the final stage, the dose-response function; the plausible range of risk estimates covers several orders of magnitude, and the linear model used by the CAG is believed by most experts to generate an upper-bound risk estimate.

The cost estimates are much less uncertain, though far from precise. The EPA states that they are "only preliminary estimates (± 30 percent)" (U.S. Environmental Protection Agency 1980, p. 5-16), but the actual uncertainties, particularly at the plant-specific level, probably are substantially greater because of the very limited number of plant-specific factors used to adjust the figures for the model plant. As suggested above, the estimates for plants that already have some form of control are likely to be too high.

In summary: The plant-specific cost and benefit estimates are highly uncertain, reflecting both general uncertainties and the use of average

Table 8.6   Cost-effectiveness of emission standards at 100 percent capacity utilization.

	Change in control level		
	From current to 97%	From current to 99%	From 97% to 99%
Annual costs and benefits			
Control cost ($10³)	2,032	2,721	689
Reduced emissions (10³ kg)	5,059	5,418	357
Reduced exposure (10³ ppb-years)	1,139	1,208	69
Reduced fatalities	0.386	0.410	0.023
Cost-effectiveness			
Emissions ($/kg)	0.40	0.50	1.92
Exposure ($/ppb-year)	1.78	2.25	10.02
Fatalities ($10⁶/life)	5.26	6.64	29.5

rather than plant-specific values for many parameters. The cost estimates, on average, may be biased upward slightly. The benefit estimates are almost certainly too high, primarily because of the "conservative" dose-response model employed. These uncertainties imply that estimates of the net benefits of any regulatory strategy will be very crude. The data do allow us, however, to compare the *relative* efficiencies of alternative strategies. Whatever the approach taken to regulating maleic anhydride plants, neither the costs nor the benefits are likely to be large. Nevertheless, analysis of alternative approaches can offer insights that are applicable to more consequential decisions.

Evaluation of the Regulatory Alternatives

Four alternative strategies will be analyzed here: a uniform emission standard, a conditional standard that sets tighter limits for plants in high-exposure areas, a uniform charge on emissions, and a uniform charge on exposure. The last alternative can also be thought of as a variable emission charge, where the charge is proportional to a plant's exposure factor.

Uniform emission standard

Table 8.6 summarizes the two standards considered by the EPA: and 97 percent and 99 percent control. The entries under "annual costs and benefits" are derived from tables 8.1–8.5 and assume full-capacity operation. The entries under "cost-effectiveness" are simply the costs

Table 8.7   Cost-effectiveness of emission standards at 56 percent capacity utilization.

	Change in control level		
	From current to 97%	From current to 99%	From 97% to 99%
Annual costs and benefits			
Control cost ($10³)	2,124	2,816	692
Reduced emissions (10³ kg)	2,846	3,048	202
Reduced exposure (10³ ppb-years)	641	679	39
Reduced fatalities	0.217	0.230	0.013
Cost-effectiveness			
Emissions ($/kg)	0.75	0.92	3.43
Exposure ($/ppb-year)	3.32	4.14	17.98
Fatalities ($10⁶/life)	9.77	12.2	52.7

divided by the relevant benefits. The modest benefits do not come cheaply; the estimated cost per life saved of the 97 percent standard proposed by the EPA is $5.3 million, an amount well in excess of the willingness-to-pay estimates cited in chapter 7. The average ratio for the 99 percent standard is only slightly higher: $6.6 million. For decision-making purposes, however, the relevant figure is the marginal ratio—the incremental cost of tightening the standard from 97 to 99 percent, divided by the incremental benefit. The higher level requires a valuation of $29.5 million per life saved.

The standards are considerably less cost-effective if the plants are not at full capacity. Table 8.7 reports the results at 56 percent capacity, the average in 1977. The costs rise slightly, the benefits fall in proportion to reduced production, and the estimated cost per life saved rises to $9.8 million for 97 percent and, on a marginal basis, to $52.7 million for 99 percent.

The high cost-effectiveness ratios for 97 percent result from the fact that, of the five plants that would need new controls, two already achieve 90 percent. For these two, the net reduction in emissions with the 97 percent standard would be very small. Thus, one potentially attractive option might be to impose a 90 percent standard, which would allow Reichold (Illinois) and U.S. Steel to use their existing controls. Although the EPA has not developed cost estimates for 90 percent, we can make a conservative estimate of the net benefits of shifting from a 97 percent to a 90 percent standard by assuming that the shift would result in no cost savings for the three plants that would still require controls, but that it would reduce the benefits from those

Table 8.8 Comparative cost-effectiveness of 90 percent and 97 percent standards for full-capacity operation.

	Change in control level		
	From current to 90%	From current to 97%	From 90% to 97%
Annual costs and benefits			
Control Costs ($10³)	1,165	2,032	867
Reduced emissions (10³ kg)	4,646	5,059	413
Reduced exposure (10³ ppb-years)	1,064	1,139	75
Reduced fatalities	0.361	0.386	0.026
Cost-effectiveness			
Emissions ($/kg)	0.25	0.40	2.10
Exposure ($/ppb-years)	1.09	1.78	11.56
Fatalities ($10⁶/life)	3.23	5.26	33.3

Table 8.9 Plant-specific cost-effectiveness ratios.

	Average cost per life[a]	
	97%	99%
Ashland	41.8	41.7
Denka	N.A.	244
Koppers	N.A.	N.A.
Monsanto	1.7	1.7
Reichold (IL)	1,882	1,074
Reichold (NJ)	N.A.	175
Tenneco	4.3	4.4
U.S. Steel	76.8	44.1

a. In millions of dollars.

plants. That is, we assume that with 90 percent, Ashland, Monsanto, and Tenneco would reduce emissions by only 90 percent, but that this would cost as much as the 97 percent controls.[8] Table 8.8 reports the results (assuming full-capacity operation). Estimated lives saved are only 6.4 percent lower, but costs are reduced 42.7 percent and the cost per life saved drops from $5.3 million to $3.2 million. More dramatically, to justify 97 percent over 90 percent, the value assigned to saving a life must be at least $33.3 million.[9]

A uniform emission standard will be least desirable when cost-effectiveness varies widely across sources. Table 8.9 presents plant-specific cost-effectiveness (lives saved) figures for both 97 percent and 99 percent. These ratios assume full capacity, and hence correspond to the aggregate figures in table 8.6. Although only five plants would be affected by a 97 percent standard, the cost-effectiveness ratios cover a wide range, from $1.7 million for Monsanto to $1.9 billion for Reichold (Illinois). The latter plant has an extraordinarily high ratio because it has the lowest exposure factor of the eight plants (so any given reduction in emissions yields minimal health benefits) and because it has already achieved 90 percent (so compliance with a 97 percent standard would have little impact on emissions).

These results suggest that a more cost-effective approach might be to make the regulations sensitive to plant-specific exposure factors or to plant-specific emission-control costs. A conditional standard, which varies with population densities, accomplishes the first goal; a uniform emission charge accomplishes the second. An exposure charge does both.

Conditional standard

The idea behind the conditional standard is to set tighter standards for firms in densely populated areas. In its extreme form it leads to plant-specific standards, since each plant has a unique exposure factor. A more practical formulation is to divide plants into a limited number of exposure classes. The small number of maleic anhydride plants makes it difficult to justify establishing more than two classes. Let us split them evenly into four high-exposure plants [Monsanto, Reichold (N.J.), Denka, and Tenneco] and four low-exposure plants [Koppers, U.S., Steel, Ashland, and Reichold (Ill.)] on the basis of the exposure factors in table 8.4.

Table 8.10 presents cost-effectiveness ratios for each group for three different standards: 90, 97, and 99 percent. In the high-exposure group, 97 percent is justified if the value per life saved exceeds $2.0 million; 99 percent is not justified unless the value exceeds $37.5 million. The cost-effectiveness ratios are even higher for the low-exposure plants at any of the three control levels. If 97 percent were imposed on the high-exposure plants, with no new controls on the low-exposure group, 96 percent of the benefits of a uniform 97 percent standard would be achieved at 37 percent of the cost. Such a two-level standard would outperform a uniform 90 percent on both dimensions, achieving slightly greater benefits at 29 percent lower cost. Thus, a simple modification of the standards approach could raise cost-effectiveness significantly.

Table 8.10   Cost-effectiveness of conditional standard.

	Control level		
	90%	97%	99%
"High-exposure" plants			
Cost ($10³)	755	755	1,398
Reduced fatalites	0.352	0.369	0.386
Marginal CE ($10⁶/life)	N.A.[a]	2.04[b]	37.5
"Low-Exposure" plants			
Cost ($10³)	410	1,277	1,323
Reduced fatalities	0.009	0.017	0.026
Marginal CE ($10⁶/life)	44.2	N.A.[a]	54.2[c]

a. Control option is nonoptimal, since higher control level yields lower cost-effectiveness ratio.
b. Relative to status quo.
c. Relative to 90% option.

Table 8.11   Plant-control combinations, in descending order of cost-effectiveness in reducing emissions.

Plant	Control %	$/kg
Monsanto	99	0.224
Ashland	99	0.238
Tenneco	97	0.285
Tenneco	99	0.383
U.S. Steel	99	2.678
Reichold (IL)	99	3.038
Denka	99	20.500
Reichold (NJ)	99	22.685

Uniform emission charge

A uniform emission charge will not be fully efficient if the marginal health damages from emissions vary across plants. It is likely to represent an improvement over a uniform standard, however, because it allocates controls in accordance with emission-control costs. As a uniform emission charge is raised, plants adopt controls in the order of their cost-effectiveness per unit of emissions controlled. Table 8.11 ranks plant-control combinations in that order. This ranking is not identical to one based on the marginal costs of controlling exposure. For example, because it is the second largest plant without controls, Ashland ranks second, after Monsanto, in cost-effectiveness of emissions control, but would rank lower in cost per unit of reduced exposure because it is in a lightly populated area. Since the two plants' estimated emission cost-effectiveness ratios differ by only 6.3 percent, it would be difficult to impose a uniform emission charge that would induce Monsanto to control without also making it advantageous for Ashland to do so. Yet, in cost-effectiveness per life saved, as shown in table 8.7, controls are far less cost-effective at Ashland than at Monsanto.

Computing the appropriate emission charge requires several steps. If the value of saving a life is V, the value of reducing exposure by one unit will be RV, where R is the risk factor. The value of reducing emissions by one unit is then RVE, where E is the number of units of exposure per unit of emissions. If we use the CAG's estimate of 0.339×10^{-6} deaths per ppb-person-year exposure, then $t = 0.339VE$, where t is measured in dollars per kg benzene emitted, V in millions of dollars, E in ppb-person-years per kg emitted. The difficulty is that E varies from plant to plant. One solution is to use the average exposure factor for E. The simple average of the exposure factors in table 8.4 is 198 ppb-person-years per thousand kg emitted. If we weight the exposure factors by plant capacity, the average is 203. For simplicity, let us use an estimate of 200 ppb-person-years per thousand kg emitted, or 0.2 ppb-person-years/kg. The average damage per kg emitted is then $t = 0.339(0.2)V = 0.0678V$. If $V = \$1$ million, the emission charge would be $\$0.0678$/kg. Since the lowest cost-effectiveness ratio in table 10.11 is $\$0.224$/kg, this charge would not induce any new controls unless $V > 0.224/0.0678 = \$3.3$ million, at which point Monsanto would control at the 99 percent level. For $V > 0.238/0.0678 = \$3.5$ million, Ashland also would add controls at the 99 percent level. Tenneco would add controls at the 97 percent level for $V > 0.285/0.0678 = \$4.2$ million and at the 99 percent level for $V > 0.383/0.0678 = \$5.6$ million. If Monsanto, Ashland, and Tenneco all controlled at the 99 percent level, the estimated benefits would exceed slightly those obtained under the uniform 97 percent standard (0.394 versus 0.386 lives saved), while costs would be 41 percent lower (\$1.2 million versus \$2.0 million per year).

These results clearly are suboptimal. Under this approach, the charge would not be high enough for Monsanto to control unless $V > \$3.3$ million, though from table 8.9 controls appear to be justified for Monsanto if $V > \$1.7$ million. Conversely, this approach leads to controls at Ashland for $V > \$3.5$ million when in fact they are not justified unless $V > \$41.7$ million. These anomalies suggest a somewhat more sophisticated approach. Table 8.9 indicates that for $V > \$1.7$ million the charge should be sufficiently small that no firm adds new controls.

For somewhat larger values of V, the charge should be high enough to induce Monsanto to control, but not Ashland. That suggests that if V = \$2 million, for example, the optimal emission charge would be in the range \$0.224–\$0.238 per kg. If, as a result, Monsanto controlled at 99 percent, 86 percent of the benefits of a 97 percent uniform standard would be realized (0.332 versus 0.386 lives saved) at only 28 percent of the cost (\$0.6 million versus \$2.0 million). However, in light of the closeness of the ratios at the two plants and the uncertainty, such a charge might lead both firms (or neither) to control.

It is clear that one would never want an emission charge that induced Ashland to control without also inducing Tenneco. If the charge were in the range \$0.338–\$2.678, both Ashland and Tenneco would control at 99 percent, with estimated incremental benefits of 0.062 lives and costs of \$649,000, for a cost-effectiveness ratio of \$10.6 million per life saved. Thus, only if $V >$ \$10.6 million should the emission charge be high enough to get any plants other than Monsanto to control.

Exposure charge
A uniform charge levied on expected damages takes account of variations in both control costs and marginal damages per unit of emissions, and leads to an optimal allocation of control efforts. If damages are proportional to exposure, as suggested by the CAG's linear dose-response function, a uniform charge on total exposure yields the same result. Although it is not possible to measure human exposure directly, estimated exposures based on measured emissions and on plant-specific exposure factors (such as those in table 8.4) could serve as the basis for an exposure charge.

The optimal exposure charge (t^*) is equal to the marginal damage per unit of exposure. In this case, $t^* = RV$, where, as before, R is the risk per unit of exposure and V is the value per life saved. Using the CAG's risk estimate of 0.339×10^{-6} deaths per ppb-person-year, the optimal charge will be $t^* = 0.339V$, where V is in millions of dollars and the charge is in dollars per ppb-person-year. For example, $t^* =$ \$0.34 if $V =$ \$1 million and $t^* =$ \$1.70 if $V =$ \$5 million. Note that the implicit *emission* charge is not uniform. If $t =$ \$1.00 (corresponding to $V =$ \$2.9 million), for example, the implicit emission charge for Reichold (New Jersey), with an exposure factor of 384 ppb-years/10^3 kg, would be \$0.384/kg. In contrast, the implicit charge for that company's plant in Illinois, which has an exposure factor of 8.13, would be only \$0.008/kg.

Table 8.12   Plant-control combinations, in descending order of marginal cost-effectiveness in reducing exposure.

Plant	Control %	$/ppb-year	Millions of dollars per life
Monsanto	99	0.57	1.7
Tenneco	97	1.46	4.3
Tenneco	99	1.96	5.8
Ashland	99	14.15	41.7
U.S. Steel	99	14.96	44.1
Reichold (NJ)	99	59.25	175
Denka	99	82.85	244
Reichold (IL)	99	365.53	1,074

Plant-control combinations are ranked by exposure cost-effectiveness in table 8.12. (If risk is proportional to exposure, this ranking is identical to one based on cost-effectiveness per life saved.) This ranking gives the predicted order in which plants would install controls as the exposure charge was raised. These ratios suggest that if the exposure charge were less than $0.57 per ppb-person-year, corresponding to $V < \$1.7$ million per life saved, no plants would add 97 percent or 99 percent controls. We may be incorrect, however, if we assume that no charge less than $0.57 would have any impact. Suppose the charge were $0.34 per ppb-person-year (implying $1 million per life saved). Given Monsanto's estimated baseline emissions of $2{,}551 \times 10^3$ kg benzene (assuming full capacity) and its exposure factor of 391 ppb-person-years per thousand kg, its total annual charges would be $2{,}551 \times 391 \times 0.34 = \$339{,}000$—an amount that might be sufficiently large to tip the balance to n-butane, as already contemplated.[10]

If the 97 and 99 percent controls are the only feasible alternatives, an exposure charge between $0.57 and $1.46 (corresponding to V between $1.7 million and $4.3 million) would induce Monsanto to control at 99 percent without affecting control efforts at any other plants. Controlling the Monsanto plant at 99 percent, however, achieves 86.0 percent of the estimated health benefits of a uniform 97 percent standard at only 27.6 percent of the cost. (Relative to a 90 percent standard, it achieves 92.0 percent of the benefits at 48.1 percent of the cost.)

Raising the charge above $1.46 would, according to the estimates in table 8.12, induce Tenneco to add controls. If the charge exceeded $1.96 (corresponding to $V = \$5.8$ million), Tenneco would achieve 99 percent control. With both Monsanto and Tenneco at 99 percent, 99.2 percent of the benefits of the 97 percent standard would be achieved at 38.6 percent of the cost. (In comparison with the 90 percent standard, 106 percent of the benefits would be achieved at 67.3 percent of the cost.) To induce plants other than Monsanto and Tenneco to control, the charge would have to be much higher—on the order of $14–$15 per ppb-person-year (corresponding to a value per life saved in excess of $40 million).

Net benefits
The preceding analyses suggest that any of the alternatives examined would be more cost-effective than the EPA's proposed uniform 97 percent standard. Each can achieve benefits of roughly the same magnitude at far lower cost. Thus far, however, we have not computed net benefits in monetary terms. For any particular value per life saved, V, the estimated annual net benefit of each scheme is the number of lives saved times V, minus costs. For each alternative except the EPA's proposed standard, the controls at each plant—and hence the costs and benefits—also depend on V. Figure 8.1 indicates which plants would add controls under each approach for values of V between $0 and $10 million. The second column, for example, shows that the "best" uniform emission standard involves no new controls for $V < $3.2 million and a 90 percent standard for $V \geq $3.2 million (necessitating controls at Ashland, Monsanto, and Tenneco).

Table 8.13 reports the net benefits for each plant for different values of V, assuming the plant-control combinations in figure 8.1. The 97 percent uniform standard yields negative net benfits for $V \leq $5.3 million. The "best" uniform standard yields positive net benefits for $V \geq $3.2 million; over the entire range $0 \leq V \leq $10.0 million, that standard's estimated net benefits exceed those of the EPA alternative by at least $600,000. The conditional standard, which imposes a 97 percent standard on only the four high-exposure plants for $V > $2.0 million, provides still higher net benefits, exceeding those of the "best" uniform standard by about $500,000 over most of the range. Columns 4 and 5 present the results for the two different versions of the uniform emission charge. Under the average version the charge is computed using the average exposure factor. That produces higher net benefits than either uniform standard, but does not do as well as the two-level conditional standard. The somewhat more sophisticated "optimal" version, which uses the exposure factors of plants at the margin, yields

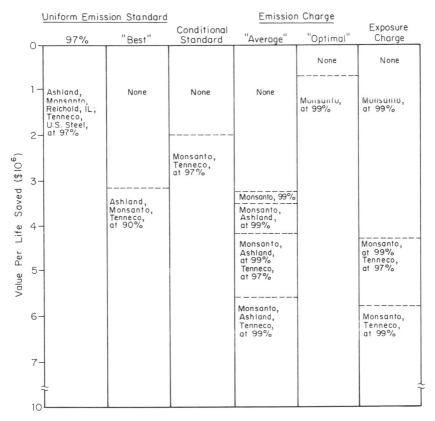

Figure 8.1 Plant-control combinations under alternative strategies.

substantial gains for smaller values of V but only modest gains for larger values. Compared with the conditional standard, it does better for $V \leq \$5.8$ million but worse for larger values.

The exposure charge dominates; none of the other approaches yields higher net benefits for any value of V. Compared with the EPA's proposal, the exposure charge generates $1.2 million to $1.8 million in additional net benefits each year, depending on the value of V. Relative to the "optimal" uniform emission charge, the exposure charge performs only slightly better, and then only for higher values of V; the maximum difference, at $V = \$10$ million, is slightly less than $0.3 million. As discussed above, however, the "optimal" emission charge is unlikely to perform as well as predicted. Compared with the "average" emission charge, the exposure charge does considerably better, with higher net benefits for all values of V greater than $1.7 million.

Table 8.13 Net benefits under alternative strategies.

V ($\$10^6$/life)	Uniform emission standard[a]		Conditional standard[a]	Uniform emission charge[a]		Exposure charge[a]
	97%	"Best"		"Average"	"Optimal"	
0.5	−1,839	0	0	0	0	0
1.0	−1,646	0	0	0	0	0
2.0	−1,259	0	0	0	+104	+104
3.0	−873	0	+353	0	+437	+437
4.0	−486	+279	+722	+385	+769	+769
5.0	−100	+640	+1,092	+761	+1,135	+1,135
6.0	+287	+1,000	+1,461	+1,153	+1,433	+1,517
7.0	+673	+1,361	+1,830	+1,544	+1,766	+1,901
8.0	+1,059	+1,722	+2,200	+1,941	+2,098	+2,284
9.0	+1,446	+2,083	+2,569	+2,335	+2,430	+2,668
10.0	+1,832	+2,444	+2,938	+2,728	+2,762	+3,052

a. Annual net benefits in thousands of dollars

Robustness under uncertainty

The net benefits presented here assume that the EPA's estimates of costs and benefits are accurate. These estimates, however, are subject to great uncertainty, so we cannot ascribe much precision to the absolute levels of net benefits. Some of these uncertainties may be resolvable once a regulation is imposed; it may be possible, for example, to obtain better plant-specific control-cost estimates. But it may then be too late to make significant changes. We may never have very firm estimates of the health benefits. A critical issue in choosing among strategies, then, is their robustness in the face of uncertainties: If costs of benefits diverge from the estimates, to what extent is a particular approach "self-correcting"?

All of the strategies are sensitive to uncertainties about the benefits. How tight an emission standard should be—whether uniform or varying with exposure—depends critically on the estimated dose-response function and on the value of improved health. These same factors determine how high a charge ought to be, whether it is a uniform emission charge or one based on exposure. To the extent that either factor is estimated incorrectly, the regulation will be set at the "wrong" level. Thus, uncertainty about damages, though it is a major problem in determining the stringency of a regulation, does not argue for one regulatory strategy over another.

In contrast, uncertainty about control costs does affect the relative attractiveness of the different approaches. If marginal damages do not vary with emissions, price-based aproaches will outperform those based on quantities when costs are uncertain. To set an appropriate standard, we needed data on costs as well as on benefits. For the uniform standard, total costs across firms were sufficient; for the two-level standard based on exposure, we needed separate costs for each group. In both cases the efficiency of the standard chosen depends on the accuracy of the cost data.

Cost data play no role in setting the uniform emission charge on the basis of average exposure. Under that approach, the charge is invariant with cost; doubling or halving the EPA's cost estimates has no effect on the charge. In contrast, the efficiency of the emission charge based on exposure in marginal plants depends heavily on accurate cost data. The results in table 8.13 for that charge assume that it is possible to establish an emission charge that would induce Monsanto to control without also getting Ashland to do so, yet the estimated cost-effectiveness ratios for controlling emissions at the two plants differ by only about 6 percent: $0.224/kg as opposed to $0.238/kg. If these cost-effectiveness ratios are accurate, an emission charge of $0.231/kg would accomplish the desired result, yielding the net benefits in table 8.13 for $2 million $\leq V \leq$ $10 million. The net benefits would be far lower, however, if the two firms' costs differed from the EPA's estimates by even modest amounts. If costs at both plants were 10 percent lower than estimated, Ashland's cost-effectiveness ratio would be $0.214/kg. This amount is less than the charge, so Ashland would control, though its cost per life saved would be in excess of $37.5 million. Suppose, instead, that costs were 10 percent higher, raising Monsanto's to $0.246/kg—more than the charge. Then neither plant would control, though control at Monsanto would still be cost-effective for $V \geq$ $2 million. Since actual costs are likely to differ from any estimates by at least 10 percent, the estimated net benefits for the "optimal" emission charge are shaky indeed.

Inaccuracies in control cost estimates have no effect on the efficiency of the exposure charge; the optimal charge is independent of costs. Changes in costs might lead to changes in control, but those changes would be appropriate. Suppose an exposure charge of $1.02 per ppb-person-year were levied, reflecting a V of $3 million under the CAG's risk estimate. Table 8.12 suggests that such a charge should induce Monsanto to control at 99 percent, with benefits of $437,000 per year (table 8.13). Now suppose that costs were only half the EPA estimates. All of the cost-effectiveness ratios in table 8.12 would be cut in half, and it would become cost-effective for Tenneco as well as Monsanto to control at 99 percent. Control efforts would be greater than predicted; but with the reduction in control costs, increased control would be optimal and net benefits would rise to $759,000.[11] With benefits linear in exposure, the appropriate exposure charge does not change even when costs are radically different.

Summary and Recommendations

A uniform emission standard fails to allocate control burdens efficiently because it fails to take account of variations across plants in either emission-control costs or damages. A high-cost plant in a lightly populated area must meet the same standard as a low-cost plant in a densely populated urban area. Both of these factors are reflected in the thousandfold difference in the estimated costs per life saved between the Reichold (Illinois) and Monsanto plants.

Uniform emission charges deal very effectively with variations in emission-control costs. Analyses of such charges usually have assumed, at least implicitly, that the damage caused by a unit of emissions is independent of the source. For some pollution problems this assumption may be reasonable; in controlling a pollutant emitted by a wide range of sources in a compact geographic region, variations in control costs would swamp variations in marginal damages. Benzene from maleic anhydride plants presents quite a different set of facts; the sources are reasonably homogeneous in production and control techniques, but are widely scattered geographically, with big differences in the health consequences of emissions. As a result, a traditional uniform emission charge may be far from optimal. In this particular case, it appears that a uniform charge would be reasonably efficient, but only because the plant with the lowest control cost per unit of emissions also happens to have the highest exposure factor.

Tighter standards for high-exposure plants represent a different approach to increasing the efficiency of regulation. The narrow differences in marginal cost-effectiveness ratios by directing control efforts where they will have the greatest impact on damages. Within any grouping of plants by exposure, however, variations in marginal costs will prevent full efficiency. The empirical results suggest, nonetheless, that a simple binary distinction between "high-exposure" and "low-exposure" plants could be much more efficient than a uniform standard.

The exposure charge is the most efficient approach, because it takes account of variations in both the marginal control costs and the marginal damages of emission. That has two desirable effects: First, it leads to an efficient allocation of control efforts among existing plants at their present locations. The net benefit estimates in table 8.13 are based on this effect alone. Second, an exposure charge creates an incentive for firms to choose plant sites where the damages are lower. This effect has not been included in the estimates because the data are insufficient. It may be important, however. Consider a firm deciding where to locate a new plant. Its preferred location, absent regulation, is next to Monsanto's plant. By choosing the Reichold location in Illinois, however, it could achieve a 98 percent reduction in damages without controlling emissions at all. Uniform emission standards or charges provide no incentive to consider that option. But an exposure charge does, because by choosing the low-damage site the firm could reduce its payments fiftyfold. Differential standards conditional on exposure factors would also affect location, though not as efficiently, nor would they be as robust in the face of uncertainty about control costs.

Postscript: The Impact of New Information

In August 1980, shortly after the preceding analysis had been completed and submitted to the EPA in draft form, the EPA held a public hearing on the proposed maleic anhydride standard. Witnesses for the Chemical Manufacturers' Association (CMA) offered information that has a significant impact on estimates of the costs and benefits.[12] This new information provides an excellent opportunity to test the robustness of the alternatives in the face of imperfect information. This postcript examines how well the different regulatory strategies, formulated with the old data, would stand up to these changes.

Witnesses at the hearing testified on a wide range of issues, including the Carcinogen Assessment Group's risk estimate. Several key points specifically concerned maleic anhydride plants: Two additional plants (Monsanto and Tenneco) will install 97 percent controls, regardless of the EPA's standard, because of state requirements and other considerations. The EPA overlooked a plant operated by Pfizer in Terre Haute, Indiana. The industry's control-cost estimates are several times higher than the EPA's. The CMA has developed exposure estimates for individual plants using an EPA-approved dispersion model but substituting plant-specific operating parameters and local meteorological conditions for the uniform values assumed by the EPA's contractor. In addition, the CMA noted that two plants, Koppers and Reichold

Table 8.14 Costs and benefits of proposed standard for the Pfizer plant.

Annual costs and benefits	
Control cost ($10³)	199
Reduction in emissions (10³ kg)	678
Reduction in exposure (10³ ppb-years)	11.1
Reduction in fatalities	0.0038
Cost-effectiveness	
Emissions ($/kg)	0.29
Exposure ($/ppb-years)	17.93
Fatalities ($10⁶/life)	52.88

(New Jersey), have closed (Galluzzo and Glassman 1980). (Both of those plants already met the proposed standard, so no costs or benefits were included for them in the original analysis.)

Changes in costs and benefits of proposed standard
The net effect of the CMA's information is to worsen substantially the estimated cost-effectiveness of the proposed standard. The controls at Monsanto and Tenneco mean that the proposed standard would not impose any costs on those plants, nor yield any benefits. Thus, the costs and benefits attributed to those plants should be subtracted from the totals. The discovery of Pfizer has just the opposite effect; it adds costs and benefits. Table 8.14 presents costs and benefits for Pfizer, using the EPA's methods.[13] Because the plant is in a lightly populated area, its exposure factor is only 16.4 ppb-person-years per thousand kilograms. Only Reichold (Illinois) has a lower exposure factor. As a result, the unit costs for exposure reduction and lives saved are high: $17.93 per ppb-person-year and $52.9 million per life.

Table 8.15 compares the original estimates of the costs and benefits of the proposed 97 percent standard (from table 8.6) with the revised estimates, after Monsanto and Tenneco have been eliminated and Pfizer added. No other changes have been made; the underlying plant-specific costs and benefits are those used by the EPA or, in the case of Pfizer, have been derived using the EPA's models and assumptions. The changes reduce the estimated costs of the proposed standard by 27 percent and decrease emissions controlled by 49 percent, thus increasing the cost-effectiveness ratio for emissions by 42.5 percent. Estimated benefits virtually disappear, however; the cost per ppb-person-year rises by a factor of 13.5, from $1.78 to $23.96, and the cost per life saved jumps to $70.7 million. The explanation for these dramatic changes is

Table 8.15 Cost-effectiveness of proposed 97 percent standard, using original and revised data.

	Original data	Revised data
Annual costs and benefits		
Control costs ($10³)	2,032	1,476
Reduced emissions (10³ kg)	5,059	2,578
Reduced exposure (10³ ppb-year)	1,139	62
Reduced fatalities	0.386	0.021
Cost-effectiveness		
Emissions ($/kg)	0.40	0.57
Exposure ($/ppb-year)	1.78	23.96
Fatalities ($10⁶/life)	5.26	70.70

Table 8.16 Comparison of EPA and CMA annual cost estimates for proposed standard.

	EPA estimate[a]	CMA estimate[b]	Increase (%)
Ashland	410	1,940	373
Pfizer	199	590	196
Reichold (IL)	320	880	175
U.S. Steel	547	2,070	278
Total	1,476	5,480	271

a. In thousands of dollars. Source: U.S. Environmental Protection Agency 1980.
b. In thousands of dollars. Source: Chemical Manufacturers' Association 1980.

simple: With the elimination of Monsanto and Tenneco and the addition of Pfizer, the standard affects only four plants. Three have very low exposure factors (and one of them already achieves 90 percent control). The fourth (U.S. Steel) has a moderate exposure factor, but it currently achieves 90 percent control, so control expenditures buy little net reduction in emissions.

The calculations above use EPA cost estimates. The CMA asked each of the firms that would be affected to estimate the cost of installing controls. (In apparent contradiction to EPA's relative cost estimates, all of the plants said they would use thermal incineration rather than carbon adsorption.) The CMA then computed the annualized capital cost using the EPA's assumptions about the appropriate interest rate and lifetime of the equipment, and added the EPA's own estimates of operating costs to obtain total annual costs (Chemical Manufacturers' Association 1980). Table 8.16 compares the EPA's estimates with the CMA's for each of the four plants that would be affected. The CMA's four-plant total of $5.48 million is nearly four times the EPA's estimate. If we substitute the CMA's cost estimate for the revised EPA estimate, the cost per ppb-person-year rises to $88, and the cost per life to $261 million.

Robustness of alternative strategies

Even before this new information, the proposed standard was difficult to justify on benefit-cost grounds. With the new information, even if one discounts entirely the higher estimates provided by the CMA, the task seems virtually impossible. Moreover, it does not appear that controls beyond those already in place or planned can be justified at any individual plant. By the EPA cost estimates, Ashland has the lowest cost-effectiveness ratios, but even these are $14 per ppb-person-year, or $41.8 million per life saved using the CAG risk estimate. If we use the CMA cost estimates, Pfizer offers the most cost-effective control opportunity, but the cost is $53 per ppb-person-year, or $155 million per life saved. Thus, it would appear that the optimal regulatory strategy would not lead to new controls at any plant, regardless of which cost estimates we believe. The maximum net benefit relative to the status quo is $0.

Suppose the EPA had implemented one of the regulatory alternatives based on the information available to it in the spring of 1980. How well would that strategy stand up to the new information? To determine the optimal levels of the alternative strategies, we need to specify a value per life saved. Suppose the EPA has used $V = \$6$ million per life saved (about $2 per ppb-person-year). (That value is high, but it suits the illustration because, when combined with the original estimates, it yields positive net benefits for the EPA's proposal.) As shown in figure 8.1, the "best" uniform emission standard would have been 90 percent, yielding estimated annual net benefits of $1 million (table 8.13).

Assume only the elimination of Monsanto and Tenneco and the addition of Pfizer. The 90 percent standard would affect only Ashland and Pfizer. Control costs would fall from $1,165,000 to $609,000, and benefits would fall from 0.361 to 0.013 lives saved. Revised net benefits, using $V = \$6,000,000$, would then be ($6 \times 10^6$)(0.013) − $609,000 = −$531,000. Thus, had the EPA had this new information, it would not have imposed the standard. Inaccurate information would have cost $531,000 per year under a uniform standard.

Suppose instead a conditional standard. With the original information, the optimal two-part standard would have been 97 percent for high-exposure plants and none for low-exposure plants. Monsanto and Tenneco would have installed controls, yielding net benefits (again assuming $V = \$6$ million) of $1.461 million. With the new information, however, no net benefits would be reaped; Monsanto and Tenneco would not be affected, and Pfizer would be in the low-exposure category. The incorrect information would not have led the EPA to the wrong conditional standard in this case; maximum net benefits ($0) would have been achieved.

Now consider the three charge schemes. Under the emission charge based on average exposure, $V = \$6$ million translates into a charge per kilogram of $0.41. The original prediction (figure 8.1) was that Monsanto and Ashland would control at 99 percent and Tenneco at 97 percent, yielding net benefits of $1.153 million per year (table 8.13). With Monsanto and Tenneco eliminated and Pfizer added, Ashland controls at 99 percent and Pfizer at 97 percent; costs are $624,000, 0.014 lives are saved, and net benefits are ($6 \times 10^6)(0.014) - \$624,000 = -\$540,000$ per year. In this case, the incorrect information costs $540,000.

With the original data, the "optimal" emission charge per kilogram was $0.224 \le t < \$0.238$ for 1.7 million $< V < \$10.6$ million, with the expectation that only Monsanto would control. For $V = \$6$ million, estimated annual net benefits were $1.433 million. With the new information, however, an emission charge in that range would have no effect; Monsanto would control regardless of the EPA regulation, and Pfizer would not find control worthwhile because its estimated cost of $0.29 per kilogram would be higher than the charge. Thus, although the "optimal" emission charge was set using incorrect information, it still leads to the optimal result.

The exposure charge also leads to the optimal outcome, with $0 in net benefits. Using the CAG risk estimate, for $V = \$6$ million, the charge is $t = (0.339 \times 10^{-6})(\$6 \times 10^6) = \$2.03$ per ppb-person-year. With the original information we predicted 99 percent controls at Monsanto and Tenneco, yielding net benefits of $1.517 million. With the new information, the charge should not have any effect; Monsanto and Tenneco already have been controlled, and Pfizer, with a cost of $17.93 per ppb-person-year, would not control. Thus, the exposure charge remains optimal, despite the change in information.

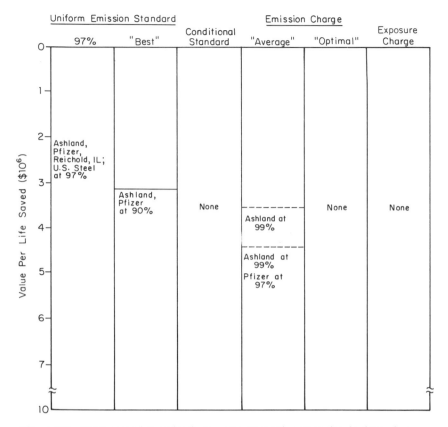

Figure 8.2 Plant-control combinations under alternative strategies, in light of new information.

Figure 8.2 summarizes the predicted effects of the different strategies given the new information on Monsanto, Tenneco, and Pfizer, but using the EPA cost estimates. Table 8.17 presents the corresponding net benefit estimates. Note the robustness of the exposure charge; regardless of the value of V, it achieves the maximum net benefits possible ($0). The optimal exposure charge, of course, is independent of the costs and existing control levels at the plants.

The conditional standard and the "optimal" uniform emission charge also perform robustly. Unlike the exposure charge, however, these strategies will not always perform so well in the face of imperfect information about costs and controls. If Pfizer already had achieved 90 percent and had an exposure factor placing it in the "high-exposure" category, under the conditional standard it would be subject to 97

Table 8.17 Estimated net benefits of alternative strategies, in light of new information.

V ($10⁶/life)	Uniform emission standard[a]		Conditional standard	Uniform emission charge[a]		Exposure charge
	97%	"Best"		"Average"	"Optimal"	
0.5	−1,466	0	0	0	0	0
1.0	−1,455	0	0	0	0	0
2.0	−1,434	0	0	0	0	0
3.0	−1,413	0	0	0	0	0
4.0	−1,392	−557	0	−384	0	0
5.0	−1,371	−544	0	−554	0	0
6.0	−1,350	−531	0	−540	0	0
7.0	−1,329	−518	0	−526	0	0
8.0	−1,308	−505	0	−516	0	0
9.0	−1,287	−492	0	−498	0	0
10.0	−1,266	−479	0	−484	0	0

a. Thousands of dollars per year.

percent for $V > \$1.7$ million. Given its preexisting controls, however, that would have been suboptimal. Conversely, suppose that Pfizer was located where it is, in a lightly populated area, but had lower emission-control costs. The "optimal" emission charge would lead it to control, though in fact such control would be suboptimal given Pfizer's low exposure. Examples such as these suggest that the apparent robustness of the conditional standard and the "optimal" emission charge in this instance is due more to luck than to inherent virtue.

As shown in table 8.17, all of the other strategies yield suboptimal results for at least some values of V. The costs of the proposed 97 percent standard exceed its benefits by at least $1.3 million over the whole range considered. It is interesting that for high values of V the uniform emission charge based on average exposure actually does slightly worse than the "best" uniform standard. Given variations across plants in the emission-exposure link, we cannot be certain that such a simple-minded charge will outperform a carefully chosen uniform standard.

Plant-specific exposure modeling

All of the calculations have employed plant-specific exposure factors based on the EPA's modeling. Those factors were based on plant-specific populations, but all of the meteorological data and plant operating characteristics (other than the emission rates) were assumed to

Table 8.18   Comparison of EPA and CMA exposure factors.

	ppb-years/10^3 kg	
	EPA	CMA
Ashland	16.8	17.6
Denka	248	156.6
Monsanto	391	148.3
Pfizer	16.4	18.8
Reichold (IL)	8.13	6.97
Tenneco	195	226.1
U.S. Steel	179	93.1
Mean (unweighted)	150.6	95.3

be identical. The CMA used a similar dispersion model and the EPA's population data, but used plant-specific data on such factors as stack height, exit velocity of the gas from the stack, and gas temperature. It also used the closest available meteorological data for each plant rather than the meteorological data from Pittsburgh that was used by the EPA for all of the plants (Galluzzo and Glassman 1980).

Table 8.18 compares the CMA and EPA exposure estimates for seven plants.[14] In four cases the CMA estimates are lower. The CMA estimate for Monsanto is less than half the EPA estimate, but that change is moot in light of Monsanto's intention to control regardless of EPA actions.

If we use the CMA's exposure factors rather than the EPA's, the estimates for the plants likely to be affected change only slightly. The reductions in exposure and fatalities decline by 12.3 percent, and the cost-effectiveness ratios shown in table 8.15 rise by roughly the same factor. The estimated net benefits in table 8.17 change even less.

Switching from the EPA's exposure factors to the CMA's would make little difference in this particular case. In general, however, the two sets of exposure factors could lead to quite different policies. The performances of all of the regulatory strategies are sensitive to the exposure estimates. The link is most obvious for the exposure charge (the charge at each plant is a function of the estimated exposure factor), but all of the regulatory alternatives share that dependence if they are to be set optimally.

The significant differences between the EPA and CMA exposure estimates suggest that plant-specific operating parameters and meteorological conditions do make enough of a difference to warrant their use, particularly if the strategy is based on an exposure charge. Moreover, the fact that the CMA was able to gather and apply the plant-specific data, and was willing to do so despite the fact that the results had little impact on the attractiveness of the proposed standard (once the Monsanto and Tenneco plants were eliminated from consideration), suggests that the costs are not excessive.

Conclusions

The new information revealed at the hearing illustrates the kinds of major uncertainties under which the EPA must make its decisions. With the addition of controls at Monsanto and Tenneco and the "discovery" of Pfizer, the estimated cost per unit of benefit rises by more than an order of magnitude. If the CMA cost estimates are correct, the cost-effectiveness ratio is 50 times higher than one would estimate using EPA's original data. The optimal emission standard is quite sensitive to such changes. In contrast, the optimal exposure charge is completely insensitive to changes in control costs or control levels at existing plants. It is also unaffected by the closing or opening (or the "discovery") of plants between the time a regulation is formulated and the time it is implemented. Moreover, the exposure charge will remain robust in the face of most changes in information or conditions that may occur in the future.

The new information forcefully raises the question whether any regulation of emissions from maleic anhydride plants is warranted. On the basis of the data currently available, it appears that no reasonable charge would have any impact on emissions. A charge based on plant-specific exposure, however, might be worth imposing. If the current estimates are correct, the charge would do no good, but neither would it do any harm (beyond some administrative costs). On the other hand, if some plants could reduce emissions by a modest amount (less than 97 percent) at low cost—an option that has not been examined by the EPA—then a charge would lead to some small reduction in health damages. That such a charge could be applied to new plants would provide an incentive for such plants to use an alternative feedstock or, if they use benzene and are located in moderate- to high-exposure areas, to install controls. An exposure charge on maleic anhydride plants also might be justified as part of a larger strategy that employed the same charge for other sources of benzene emissions.

Chapter 9

The Case for Exposure Charges

Charges offer a feasible and efficient alternative to standards for regulating benzene from maleic anhydride plants. The present study also suggests that the conceptual focus needs to be shifted from emissions to exposure and damages. That emissions matter only because they cause damages is obvious, but easily forgotten by those who study regulation and by those who regulate. This point is not a quibble; it implies major changes in regulatory strategies, whether they employ the charges favored by most economists or the standards favored by most legislators and regulators.

One must always be careful in attempting to generalize from a single case study, particularly one as narrow as the preceding chapter. Even readers who accept the exposure-charge concept for maleic anhydride plants may doubt its general usefulness. In particular, two major objections may be raised: that variability in the emission-exposure link, though always present, is usually less critical; and that exposure charges are impractical when the regulator must deal with many small sources rather than a handful of large plants. This chapter attempts to meet these objections and others—in part by showing how exposure charges could be applied to other source categories—and to suggest how a general strategy based on exposure charges could be structured.

Alternatives for Service Stations and Automobiles

Variability in marginal damages is even greater for automobiles and service stations than for maleic anhydride plants. This should not be surprising, in view of the wide distribution of both types of sources. Automobiles are driven along remote dirt roads in Alaska as well as on the crowded streets of Manhattan, and service stations are found in a similar range of locations. As a result, the damage caused by a kilogram of benzene emitted from an automobile or a service station varies by a factor of more than 100—estimated exposure per kilogram is 164 times higher in New York than in the Wichita standard metropolitan statistical area, for example (Nichols 1981b, chapter 11). These variations suggest the desirability of strategies that are sensitive to differences in marginal damages. In light of the large number of automobiles and service stations, however, strategies based on direct, continuous monitoring of emissions obviously are infeasible, whether they employ standards or charges. (For a more extensive analysis of the costs and benefits of regulating benzene from service stations and automobiles, see chapter 11 of Nichols 1981b.)

Service stations ✳

Benzene emissions from service stations occur primarily when the stations' underground tanks are being filled by delivery trucks ("stage I") and when individual vehicles are being filled ("stage II"). Stage I controls are relatively inexpensive, and are already required to meet state hydrocarbon standards in more than a dozen air quality control regions (AQCRs) encompassing major metropolitan areas. Stage II controls are more expensive and have been installed only in the San Francisco and San Diego areas (Mara and Lee 1978, p. 105). Thus, current regulation already incorporates a degree of geographic variability.

To impose an exposure charge requires estimation of each station's exposure factor and emissions. The large number and small size of service stations precludes the source-specific estimates of exposure factors possible with maleic anhydride plants and other large sources. However, exposure factors could be estimated for a limited set of categories defined by population densities and meteorological conditions; all of the stations in a given category would be assigned the same exposure factor.

Emissions can be estimated indirectly using three variables: the type of controls installed (if any), the benzene content of the gasoline, and the number of gallons pumped. For convenience, the charge could be collected from wholesale gasoline distributors rather than directly from the stations; this would reduce drastically the number of transactions for the agency. The charge on distributors might be based on the assumption that none of the outlets were controlled; individual stations that installed controls could then apply for rebates, with appropriate documentation showing the type of control installed and the amount of gasoline pumped. This approach offers significant advantages over either a standard or a uniform emission charge. Unlike a geographically variable standard, it takes account of variations in emission-control costs. Unlike a uniform emission charge, it takes account of variations in the damage caused by emissions.

Preliminary cost-effectiveness calculations suggest that the benefits of controlling benzene *per se* from service stations are minimal relative to the costs, even in densely populated cities. Installing stage II controls at an average-size station in New York City, for example, would reduce annual benzene exposures by less than 100 ppb-person-years, yet would cost about $900 per year (Nichols 1981b, p. 354). Gasoline vapors, however, contain many harmful substances besides benzene; thus, controls may be cost-beneficial, at least for some stations in some locations. The exposure-charge strategy described above could be used to control exposure to gasoline vapors generally, with the charge calculated to measure the damages caused by all harmful components.

Automobiles
As with service stations, automobile regulation already incorporates a limited degree of geographic diversity; automobiles in California must meet tighter emission standards than those in other states.[1] That concept could be refined, either by increasing the number of categories or by redefining the two that now exist. Harrison (1975) has shown, for example, that a "two-car" strategy that maintained tight standards for 43 large, dense SMSAs, but relaxed them elsewhere, would yield significant cost savings with very little reduction in benefits.

An exposure-based charge could be levied on new automobiles at the time of sale; the charge would be a function of that model's emissions and the exposure factor for the region in which the car was sold. (To avoid "bootlegging," it might be desirable to base the charge on the location where it was first registered rather than sold.) Such a charge would not entail significantly greater administrative burdens than the current standards.

Several additional refinements should be considered. One would be to make the charge annual rather than one-time; that would simplify the handling of cars that are moved from one place to another and also provide an incentive for owners to junk high-emission cars earlier. Such a charge could be collected along with the excise taxes levied on automobiles by many states; each vehicle's identification number could include a code indicating its emission category. A related refinement would be to base the charge on periodic emission tests; this would encourage owners to maintain emission-control devices. Another desirable refinement, if tamper-proof odometers became available at reasonable cost, would be to base the charge on miles driven; that would provide an incentive to drive less and a differential incentive for those who drive a great deal to purchase automobiles with low emissions.

As with service stations, crude cost-effectiveness calculations suggest that the benefits of controlling benzene emissions from automobiles almost certainly would be far lower than the costs, even in densely populated cities, in light of the high levels of control already achieved under existing regulations. In Chicago, for example, a car meeting current hydrocarbon emission standards causes only about 1.5 ppb-person-years of exposure to benzene per year (Nichols 1981b, p. 361). An exposure-based charge strategy of the type discussed above, however, would appear to offer a viable and efficient alternative to existing uniform standards for a variety of pollutants, though a detailed analysis of that option is beyond the scope of this study.

Exposure Charges and Efficiency

Economists and others have long criticized uniform standards for failing to take account of variations in control costs. Uniform emission charges deal with such variations automatically; a single charge allocates control efforts differentially according to costs. Such charges fail to take account of variations in the damages caused by emissions, however, and thus do not allocate control efforts efficiently. Where these variations are great—as in the cases of service stations, automobiles, and maleic anhydride plants—emission charges are likely to be far from optimal, even though they may yield substantial gains over uniform standards.[2]

Under a uniform exposure charge, each source controls to the point where its marginal cost of controlling exposure is equal to the charge. The marginal cost of controlling exposure is a function of two variables: the marginal cost of controlling emissions divided by the exposure per unit of emissions. If expected damages are proportional to exposure, as appears to be the case with benzene and other environmental carcinogens, the resulting allocation will minimize the cost of achieving any given reduction in damages. In contrast, a uniform emission charge minimizes the cost of achieving a given reduction in emissions; costs could be reduced with no change in damages by tightening controls at high-exposure sources and loosening them at low-exposure sources.

Another advantage of an exposure charge is that it provides an incentive for firms to select low-damage sites. In essence, it opens up another technique for reducing damages. The importance of this option will vary among source categories. It is likely to be relatively unimportant for such sources as automobiles and service stations, for which locational flexibility is minimal. For some sources, however, such as large chemical plants producing hazardous substances, it may be less expensive to choose a low-damage site than to control emissions. For very hazardous activities, both options may be desirable. Toxic-waste disposal sites and nuclear power plants, for example, are not permitted in major cities and must follow strict safeguards.

Exposure Charges and Equity

Throughout this study, the decision criterion has been the maximizing of aggregate net benefits. Regulatory strategies affect the distributions of costs and benefits, however, as well as their totals. When initially presented with the idea of exposure-based charges or standards, many individuals raise a variety of equity-based objections.

The most common objection arises when the exposure charge is stated in terms of differential charges on emissions depending on location. These differentials may appear to discriminate against people who live in lightly populated rural areas, placing a lower value on protecting their health. In fact, however, a uniform exposure charge — which is equivalent to emission charges that vary in proportion to exposure factors — is based on the premise that equal weight should be attached to protecting all individuals, regardless of where they live. In contrast, a uniform emission charge (or standard) implicitly assigns radically different weights. If one plant has an exposure factor 100 times that of another, a uniform emission charge implies that protecting an individual living near the first plant is worth only 1 percent as much as protecting an individual near the second.

Another possible measure of equity is the extent to which risks are distributed reasonably equally and are not concentrated among a few individuals. This criterion cannot be applied rigorously, as exposures and individual susceptibilities will vary regardless of the regulatory strategy. It certainly cannot be applied on a source-by-source basis. Whatever the validity of this criterion, in many cases exposure charges will lead to less variability in risk than uniform emission charges or standards. If a uniform emission standard is imposed, emissions per unit of production will be constant but ambient concentrations will vary widely depending on the numbers and sizes of sources. Because many sources (particularly automobiles) tend to be located where people are, those who live in densely populated areas will face higher exposures. An exposure charge, by inducing tighter controls on sources in high-damage areas, will tend to offset this effect, leading to less variation in exposure.

Still another potential objection to exposure charges is that they would have differential impacts on otherwise identical firms depending on their locations. Coupled with that objection may be the fear that high-exposure urban areas will be disadvantaged in attracting and retaining industry. That exposure charges would have differential effects on firms in different areas, and that these differentials would affect at least some location decisions, is undeniable. Indeed, these differentials are key to the efficiency of exposure charges. But no form of regulation affects all firms equally, and it is not clear why it would be any more unfair for firms in high-damage areas to pay more for emissions than it is for a firm in New York City to pay more for the land it rents. In both cases the prices reflect the opportunity costs. And just as differential land rents lead firms needing large sites to locate where land prices are lower, so too would an exposure charge lead some firms emitting hazardous substances to locate where marginal damages are lower.

Structuring the Charge

Regulators are likely to question the practicality of exposure charges on two grounds: With the tremendous uncertainties and disagreements about dose-response relationships and the dollar values of reducing risks, how can the charge rate be set? How can compliance be monitored and enforced?

Setting a specific charge would be difficult. Whatever the substance, the scientific uncertainties are likely to be great, and placing a dollar value on protecting health raises sensitive political and ethical issues. Scientists can help in narrowing risk estimates, and empirical estimates of willingness to pay for risk reduction provide some guidance. The analysis in chapter 7, for example, suggests that it is difficult to justify a charge for benzene in excess of $1 per person, and that a charge in the neighborhood of $.10 would be more reasonable. Ultimately, however, the charge cannot be chosen on purely technical grounds, but must be based in large part on the political and ethical judgments of responsible officials. The best we can hope to do is to get it roughly right.

These difficult decisions cannot be avoided, though they may be obscured, by using standards. If standards are to be set efficiently, judgments must be made (at least implicitly) about the magnitudes of risks and the value of reducing them. Uncertainties about damages are always a problem; they pose the same obstacles to setting the right standard as to setting the right charge. If standards are used, regulators must grapple with these issues many times for each substance, because each category of sources is subject to a separate standards. Moreover, control costs must be estimated for each category. In contrast, a regulator need set only one exposure charge for each substance, and the appropriate charge is independent of control costs.[3] The charge should be indexed to avoid the erosion of its effect due to inflation.

Applying an exposure charge to individual sources requires only a moderate amount of ingenuity. We cannot attach monitors to each individual and monitor exposure directly, but the link between exposure and monitorable characteristics can be estimated for individual sources or for groups of sources. For large stationary sources, such as maleic anhydride plants, source-specific dispersion modeling and population data can be used to estimate exposure per unit of emissions. Emissions can then be monitored and used to generate exposure estimates on which the charge can be levied.

For smaller sources, the approach has to be cruder and less direct. Exposure factors can be estimated for different areas. Where emissions per se cannot be monitored, they can be estimated from other data, such as control equipment and number of gallons of gasoline pumped in the case of service stations. The ongoing monitoring requirements need be no more complex than with standards.

The exposure charge has the virtue that it is relatively easy to extend to additional substances consistently, equitably, and efficiently. Setting the charge would be difficult the first time, but subsequent charges could be based on relative potencies, which are subject to less dispute than estimates of absolute risks at low doses. If a charge were set for benzene, and the next substance's estimated potency were twice that of benzene, its exposure charge would be twice as high. Consistency in relative charges could be achieved without fully resolving the disputes about low-dose extrapolation and the value of risk reduction. Such a strategy would be more efficient and could be implemented more swiftly than a "generic" policy based on standards, such as that proposed by the EPA.

Case Study 3

Air Quality Under the Clean Air Act
Robert Repetto

Chapter 10

Prevention of Significant Deterioration:
Problems and Regulations

In the graphs in economics textbooks, where marginal pollution damages accelerate smoothly as emissions increase and marginal costs of abatement accelerate smoothly as emissions are reduced, efficient regulatory policies hold emissions at the level at which the marginal damages equal marginal costs, because that minimizes the total burden. Since both the damage function and the abatement-cost curve are assumed to be known, the policy-making agency can achieve this efficiency by setting emissions standards or emission fees or by establishing an overall limit on emissions and allocating the rights to emit among polluters by a system of buying and selling entitlements. All these methods minimize abatement costs and ensure that those costs do not exceed benefits.

Compared with the actual problems facing regulators of air pollution, this representation of the problem is so idealized that it could be construed as a mockery. In the first place, many of the activities that place offensive matter in the atmosphere do so at rates that can be measured with difficulty, if at all. Emissions from factory or utility smokestacks can be monitored, although keeping track of the hundreds of trace elements that may be present (depending on the composition of the fuel) is difficult. Emissions from motor vehicles depend on the temperature, the length of time since each engine's last tuneup, and other variables. Emissions from a city's tens of thousands of furnaces and woodstoves can hardly be measured. In some industrial establishments, gases escape from scores of cracks, joints, and vents. Particles are thrown into the air in the yard while materials are being stored or shifted, or by the comings and goings of trucks over unpaved surfaces. The difficulty of measuring emissions within reasonable limits of accuracy and cost obviously impedes efforts to tax or enforce private property rights in them.

When emissions rates are hard to measure, oftentimes the costs of control measures can be calculated; but their effects remain uncertain. In many other circumstances, the cost of abatement is known only approximately by the regulatory agency. The cost and efficiency of control equipment retrofitted to existing facilities or incorporated into a new installation will depend on the scale, the remaining service lifetime, the design, and the layout of the plant in question, and on the way the control equipment and the plant are operated and maintained. Much of this information cannot be known accurately by the regulatory agency for all the many sources of emissions under its jurisdiction. The individual emissions sources, at least in industry, tend to have much more exact knowledge of their abatement cost schedules than does the agency.

In a textbook graph, damages are depicted as dependent on emissions. In fact, they are dependent on the concentration of the offending substances in the air. Predicting the effects of known emissions on air quality is a formidable undertaking (APCA 1980). The movement of particles or a gas in the atmosphere is influenced by its composition, its velocity, its temperature relative to the surrounding air, the speed and direction of the wind horizontally and vertically, contact with terrain and structures, and other factors. Accurate prediction would require complex numerical models and extensive data.

Models in use are badly misspecified and tend to be gross oversimplifications. Such misspecification can lead to large prediction errors (SAI 1975). Tests of models have shown that the size and direction of errors are not uniform, but depend on the distance from the source of emissions, the atmospheric stability and wind speed, the nature of the terrain, and the level of concentration. Within 20 kilometers of the source, specification bias can easily be a factor of 4 in either direction. More elaborate models may reduce specification error, but are much more demanding of input data and computation. Larger errors in constructed or interpolated input data may offset such models' greater theoretical accuracy. The predictions generated by different models and procedures can vary widely (Turner 1979). Few models have been tested adequately over a wide range of conditions. Thus, there is considerable uncertainty regarding the bias in any model's predictions (U.S. EPA 1978).

Data availability usually leaves much to be desired. For new sources, prediction of emissions rates is difficult. Meteorological records over periods of time for a specific area are often unavailable. Often, inferences are made from short historical records—perhaps those of only a single year—which might be quite unrepresentative of the usual weather patterns. Finally, the measurements on record could be incomplete, especially with regard to winds at elevations above surface level, which are not easily measured and typically correspond poorly to surface flows. Yet, winds at higher elevations markedly affect the transport of pollutants.

There is considerable uncertainty about the impact of given emissions on air quality, especially where the weather patterns are variable and the underlying terrain is rough. This is true for stable pollutants, which are not transformed chemically in the air. But smog, acid rain, and other air-pollution problems result from chemical reactions in the atmosphere. These transformations are highly complex sequences of reactions whose rates depend on the relative concentrations of precursors, the degree of mixing in the air, the amount of sunlight and water vapor present in the atmoshere, and the presence of catalytic trace elements (U.S. EPA 1980). Atmospheric transformations take place over hours or days, so secondarily formed pollutants appear at considerable distances from the emission source. Acid rain from industries south of the Great Lakes is destroying fish populations in the lakes of Ontario, and haze created mainly by sulfate particles originating in the smelters of Arizona is lowering visibility in Utah, New Mexico, and Colorado. These regional effects depend also, in part, on the chances of precipitation and on large-scale weather patterns. All this contributes to the difficulty of linking emission rates to ambient concentrations of pollutants.

In attempting to assess and limit pollution damages, the question also arises which pollutants to regulate. Of the thousands of substances emitted into the atmosphere, current regulations govern only a small number. These are not necessarily the specific substances to which most damages can be attributed. For example, concentrations of total suspended particulates (TSP), measured in micrograms per cubic meter, are governed by standards. In any atmosphere there will be a distribution of particles by size and chemical composition. Small particles are more damaging to health and aesthetic values (Ahmed and Perera 1978). They penetrate deeply into the lungs, avoiding natural filtration, and they scatter light more efficiently, reducing visibility much more efficiently than the same mass of large particles would. Some kinds of particles of a given size are much more dangerous than others, because they contain carcinogenic or otherwise highly toxic compounds. The difficulty of isolating and controlling the specific substances responsible for damages makes the correspondence between ambient-air-quality standards and pollution damages rather loose.

These complications seem formidable, but one of the worst obstacles to the implementation of the economists' simple prescription is the problem of predicting the damages once atmospheric concentrations of pollutants are known (U.S. EPA 1981b). Part of the problem is that the possible damages are so diverse: reducing visibility across the Grand Canyon, reducing crop yields in the Imperial Valley, increasing the weathering of the marble in the Taj Mahal, reducing the numbers of trout caught by fishermen in the Adirondacks, increasing the number of times per year our suits have to be sent to the cleaner, increasing the mean temperature of the planet. The direct health effects alone could range from increased numbers of runny noses to increased numbers of attacks among asthma sufferers, to increased mortality rates from cancer and heart disease. The marginal damages from a deterioration in air quality would be the sum of damages of many diverse kinds.

Estimates of each kind of marginal damage have to be constructed by establishing a dose-response relationship, estimating the numbers of individuals exposed to the pollution, and finding a means of evaluating the damage in terms commensurate with other kinds of damage and with abatement costs. This can be done only very imperfectly. Dose-response functions are hard to establish. Controlled experiments obviously cannot be made in which large human populations are exposed to low dosages over long periods of time and then followed up for delayed reactions. Experiments involving small nonhuman populations exposed to high dosages over short periods of time are a poor substitute. Extrapolation of damages observed at high concentrations to low-dose effects is without firm theoretical or empirical foundation (Ferris 1978). Statistical estimation of dose-response relationships from epidemiological and related data is difficult (Lave and Seskin 1977). The concentrations of pollutants tend to vary together and to be correlated with characteristics of the communities that produce them. The appearance of damages may lag well behind the episode of pollution in time. The damages from different pollutants present in combination may be not additive, but interactive. Moreover, dose-response functions describing health effects are likely to be specific to individuals of given age, health status, and genetic susceptibility, so that the aggregate dose-response function for any population depends on the mix of individuals in that population and cannot be extrapolated to another population with a different composition.

The valuation of different kinds of damages in dollars, once damages are estimated, is no less difficult. There are serious problems with all the methodologies advanced for constructing these critically important estimates (Freeman 1979). Inferences of the value of damages from market behavior, such as the study of property-value differentials as indicators of the valuation of cleaner air or the study of wage differentials as indicators of the valuation of lower occupational mortality or morbidity risk, are valid only under restrictive assumptions. Moreover, only with great difficulty can such studies disentangle the effects of the environmental factor from those of other factors correlated with it. The alternative approach, since people do not have the opportunity to buy freedom from exposure to air pollution, is to ask what they would pay if they had the chance. Responses to such hypothetical questions are not necessarily reliable, and might even be systematically biased if people thought their reponses would actually result in their having to contribute what they say cleaner air is worth to them (since they could not be excluded from enjoying the benefits of cleaner air, no matter what they contributed).

In short, the quest for greater efficiency in the regulation of air pollution is a search among areas of great uncertainty. The problem in the use of economic incentives is not finding the point on known functions at which marginal costs equal marginal damages. It is, in large part, the difficulty of designing regulatory systems to provide incentives that elicit informed decisions based on benefit-cost comparisons, and of having this done by the entities best able to obtain the relevant information.

Against this background, the regulatory strategy reflected in the Clean Air Act[1] appears as an effort to evolve a pragmatic, enforceable policy. It is a collection of regulatory programs to deal with different sources and kinds of air pollution. It certainly cannot result in economic efficiency through balancing benefits and costs, or even the more limited goal of cost-effectiveness. Recent innovations in policy have been attempts at modifying regulations to reduce excess costs.

The Environmental Protection Agency is required to set and review national uniform ambient air quality standards.[2] This has been done for the "criteria pollutants": sulfur dioxide, nitrogen oxides, carbon monoxide, ozone, hydrocarbons, and total suspended particulates. An ambient standard has recently been proposed for atmospheric lead. Ambient standards are average annual concentration and short-term average concentrations not to be exceeded more than once a year. The primary standard is set at a level designed to protect the public health with an adequate margin of safety. The more stringent secondary standard is set to protect the general welfare from all known or anticipated adverse effects. This legislative definition of *standard* severely limits the EPA's role in balancing benefits against abatement costs.

In addition, the EPA is authorized to set emissions standards for new facilities in certain important categories of stationary sources.[3] These standards specify quantitative limits reflecting the best demonstrated control system, considering the costs of abatement. The EPA must set strict standards, providing an ample margin of safety, for sources that emit especially hazardous substances. In the case of the carcinogen vinyl chloride, for example, this requirement has led to a zero emissions goal. For mobile sources, technology-forcing emissions limits were built into the 1970 legislation.[4] Because the automotive industry declined to be forced (Krier and Ursin 1979), weaker limits and postponed attainment deadlines were adopted in the 1977 amendments.[5]

Primary responsibility for the attainment of these goals and the enforcement of standards was placed in the hands of state regulatory agencies.[6] Each state was required to submit to the EPA an implementation plan to ensure attainment of ambient-air-quality goals by 1975. The states were empowered to use a wide variety of means, including comprehensive transport control programs for mobile sources, emissions standards for stationary sources, and incentive and disincentive measures. The formulation and approval of State Implementation Plans (SIPs) went slowly, and the plans were liberal with waivers and exemptions to protect emitters from economic disruption (Walker and Storper 1978). Few states showed enthusiasm for the more extreme measures (particularly to reduce automotive pollution) envisaged by the 1970 Act (Grad et al. 1975). Partly as a result, air quality in 1975, and indeed in 1979, remained in violation of national standards in many areas.

These implementation problems have given rise to two major new regulatory programs, introduced by the EPA after the passage of the 1970 Amendments and given legislative standing in 1977. The first of these arose from the nonattainment of air-quality standards by 1975 in many areas. Conceivably, all new sources of emissions could have been prohibited for contributing to the violation of a standard. However, the EPA issued an interpretive ruling[7] permitting large new stationary sources to locate where they would affect air quality in nonattainment areas provided that the source meet the "lowest achievable emissions rate" (LAER) demonstrated for that source category, that all facilities in the region controlled by the source be in compliance with all air-quality standards, and that any emissions by the new source be offset by reductions elsewhere, so that air quality would improve to maintain reasonable further progress toward attainment of the National Ambient Air Quality Standard. This approach is truly innovative, since it implies that emitters have a property right in their emissions that is transferable to other sources (Yandle 1978). It opens the possibility that new sources that could eliminate their remaining emissions after adoption of LAER only at great cost, if at all, could pay other local sources that could more cheaply make offsetting reductions to do so. This obviously leads in the direction of a market in emissions rights, in which those sources with relatively high abatement costs would appear as buyers and those with low abatement costs as sellers (Liroff 1979). Such a market would promote, if not the equation of marginal pollution damages with marginal abatement costs, at least the equation of one source's marginal abatement costs with those of others. In principle this was an important regulatory innovation to promote cost-effectiveness, but it has had little practical impact, since trades are difficult and are permitted only after adoption of LAER controls. The 1977 Amendments also put off the deadline for attainment of air standards nationwide until 1982 for stationary sources, and gave states until 1979 to submit implementation plans to achieve this.[8] States had the options of retaining the offset market or ensuring sufficient abatement by other means to provide a margin for industrial growth without endangering progress toward attainment. Most states have chosen to retain the offset provision, even though few new emissions sources have purchased emissions rights from other firms yet (Dames and Moore 1980). New sources wishing to locate in nonattainment areas have typically either reduced their planned emissions to squeeze under the minimum size exemption from the offset requirement (100 tons per year of emissions), or obtained offsets by reducing emissions from other facilities under their own control. However, further EPA policy modifications permitting sources to "bank" emissions rights by securing emissions reductions in the present to offset new emissions in the future should facilitate interfirm transactions in emissions rights.[9]

A related regulatory program that is the focus of this study governs the protection of air quality in areas where it surpasses national standards.[10] This has been one of the most controversial aspects of air-pollution regulation. It virtually negates the idea that air-quality standards balance benefits and costs, by denying that at air-quality levels superior to standards the benefits of abatement are less than the costs. The Prevention of Significant Deterioration (PSD) regulations have been the object of continual debate, litigation, reconsideration, and revision since their inception in 1970 (Stern 1977).

One of the purposes of the Clean Air Act Amendments of 1970, as of preceding legislation, was "to protect and enhance the quality of the nation's air resources" (U.S.C. 7401 b [1]). This was intended to authorize a national program to prevent degradation of air quality. The Senate report stated: "In areas where current air pollution levels are already equal to or better than the air quality goals, the Secretary shall not approve any implementation plan which does not provide, to the maximum extent practicable, for the continued maintenance of such ambient air quality."[11] The House report concurred.

When it appeared that the EPA would approve proposed State Implementation Plans that would permit growth in attainment areas which would lead to additional pollution within the NAAQS, the Sierra Club and associated groups sued to enjoin the EPA from granting approval.[12] The U.S. District Court in the District of Columbia granted an injunction. Eventually, on a 4–4 split, the U.S. Supreme Court allowed the lower court's ruling to stand, forcing the EPA to disapprove SIPs that did not contain measures to prevent the degradation of air quality and to propose regulations with which the SIPs should comply.

The EPA then aired some of its difficulties in proposing PSD regulations, suggested a half-dozen alternative approaches, and called for public comments.[13] One of the alternatives was a system of emissions charges, which the EPA discarded because it lacked the basis to associate a charge level with a target level of air quality and because the charge system could not guarantee the prevention of significant deterioration. Another discarded alternative was a system of area-wide emissions density limitations. A third was a procedure for local determination whether a particular proposed new source's emissions would result in significant deterioration, on an incremental basis.

The EPA finally adopted ambient limits to changes in air quality and zones of more and less stringent protection.[14] It set numerical limits to ambient deterioration for regulated pollutants in each zone, the most stringent protecting national parks, wilderness areas, and other regions of special scenic importance. State Implementation Plans were required to include measures to safeguard these limits of deterioration.

In addition, the PSD policy required preconstruction review of major new sources and modifications of existing sources, including monitoring of air quality before and after construction, modeling of the air-quality impacts of the proposed source to preclude infringement on the ambient limits, and emissions controls that would reflect the "best available control technology" for the type of source under review.

This basic approach, although much modified and revised, has withstood legal challenges from environmental and industrial groups and was included virtually intact in the Clean Air Act Amendments of 1977.[15] Responsibility for implementation of the legislation was initially assumed by the EPA, but was to be taken over by state agencies by March 1979. As of January 1981, only 19 states had approved PSD regulations in their SIPs or had been delegated authority by the EPA to administer PSD regulations. One important reason for this delay is the uncertainty that was created by further legal challenges to the EPA's regulations for implementation of the 1977 amendments.[16] The following paragraphs describe in more detail the current regulations promulgated on August 7, 1980, which embody revisions necessitated by the latest court findings.[17]

The Act designates mandatory Class I areas, including all international parks, wilderness areas, and memorial parks larger than 5,000 acres, and national parks larger than 6,000 acres. Other natural areas, including national preserves, monuments, and recreation and seashore areas, if originally designated as Class II areas, could only be redesignated as Class I. Other Class II areas, which originally included all other attainment areas, could be redesignated as Class III to relax ambient standards only after public hearings and approval from appropriate authorities.

Maximum allowable increases in pollutant concentrations above the baseline concentrations were defined for TSP and SO_2, with the constraint that total concentrations could not exceed national primary or secondary standards. The baseline concentration is now defined for each regulated pollutant as that experienced in the area of impact at the first date after August 7, 1977 (the date of passage of the CAAA) on which a completed PSD application is submitted. The legal increments are shown in table 10.1. Short-term increments can be exceeded only once per year, on the average. The Act also requires the EPA to study and formulate regulations governing hydrocarbons, CO, NO_x, photochemical oxidants, and other "criteria pollutants" within two years of the date of passage. These regulations were not required to involve

Table 10.1 Allowable increments in ambient concentrations in PSD regions.

	$\mu g/m^3$ [a]	
	Particulates	SO_2
Class I areas		
Annual geometric mean	5	
Annual arithmetic mean		2
24-hour maximum	10	5
3-hour maximum		25
Class II areas		
Annual geometric mean	19	
Annual arithmetic mean		20
24-hour maximum	37	91
3-hour maximum		512
Class III areas		
Annual geometric mean	37	
Annual arithmetic mean		40
24-hour maximum	75	182
3-hour maximum		700

a. Micrograms per cubic meter.

ambient increments if some other equally effective PSD strategy were to be adopted instead. Because of impending revision of the Clean Air Act, the EPA has not published such regulations.

The preconstruction review process for major new sources and modifications of existing sources was defined to include three components:

• demonstration (through the modeling of air-quality impacts of the source together with all other applicable emissions changes from other sources in the area occurring after the baseline date) that the facility will not lead to violations of any increment or other applicable national standard, or affect air-quality-related values in Class I areas, or affect air quality in any nonattainment area,

• assurance that the source will employ the "best available control technology" (defined on a case-by-case basis as the emissions limitation no less stringent than the applicable New Source Performance Standard that would achieve the maximum feasible emissions reduction, taking into consideration energy, environmental, economic, and other costs), and

• completion of the required preconstruction and postconstruction ambient air quality monitoring.

"Major" new sources are defined as those in any of the 28 specified source categories with a maximum capacity to emit any regulated pollutant under applicable emissions standards and permit conditions greater than 100 tons per year, or any other source that could emit 250 tons per year of a regulated pollutant under applicable emissions controls. A "major" modification of an existing source is defined as one that results in new increases in emissions of any regulated pollutant greater than specified *de minimis* amounts, which are 40 and 25 tons per year for SO_2 and particulates respectively.

In the 1977 Amendments, Congress particularly sought to protect and enhance visibility in Class I areas,[18] by directing the EPA to promulgate guiding regulations and to require states to incorporate measures that would ensure reasonable progress toward this goal in their SIPs. These measures include provisions for large existing stationary sources that impair visibility to install the "best available retrofit technology" (BART).

The EPA has identified 156 Class I areas in which good visibility is necessary and must be protected. As a first step, the EPA promulgated regulations requiring each state to analyze the impacts and control possibilities of major existing stationary sources and clusters of sources that can be determined to reduce visibility in these protected areas.[19] These analyses could lead to the imposition of BART. When more is known about the long-range effects of regional emissions on visibility, these regulations may be broadened. For major new sources and modifications, the EPA regulations require that visibility review be integrated with PSD preconstruction review, and that the states deny permits to major new sources that would have adverse impacts on visibility in protected Class I areas.

The Clean Air Act Amendments and subsequent EPA regulations make it clear that the specified increments and air-quality-related values have to be protected. States that find violations of increments have to take remedial action, possibly including additional control on existing sources. Consequently, rights to pollute are definitely limited. However, the regulations did not directly address the question of how entitlements to increase emissions in clean-air areas within the limitations imposed by the legislation were to be assigned. The EPA has been awarding the rights to "consume the increment" by increasing emissions and ambient concentrations above the baseline amounts on a "first-come, first-served" basis. That is, the potential new source that had first completed its application, should it comply with the conditions of the new-source review, would be awarded *gratis* the right to increase the ambient concentrations in the area of its facility by any amount within the defined increments.

When publishing its regulations, the EPA recognized that this practice might not be an efficient or equitable way of allocating a potentially scarce and valuable right, and that it denied to local communities any authority over the number, nature, or identity of the potential new sources that would be entitled to these rights. Consequently, in the EPA regulations regarding approval of State Implementation Plans for PSD[20] it was stated that SIPs should contain provisions for both the protection and the allocation of these entitlements. The following passage speaks directly to that issue:

The Administrator is concerned that while States are developing their own PSD regulations and the EPA is administering the PSD program, EPA should not make decisions which would have a significant impact upon the future growth options of the State. In the interim, EPA generally will allocate use of the increments on a first-come, first-served basis as has been done under the previous PSD regulations. The administrator recognized that this approach may not be adequate on a long-term basis to achieve the purposes of the Act. Other options are available and should be pursued by the States in the development of their plans for PSD. . . . States are required to develop a program for increment allocation and a number of program options are suggested for their consideration. EPA will be assessing the merits of and feasibility of several allocation options (including first-come, first-served) and thereafter (will) issue guidance for the submission of State implementation plans. The evaluation will consider alternatives in which carefully designed economic incentives serve as an adjunct to or a replacement for an administrative permitting procedure. The economic incentive programs to be considered include marketable permits, emission fees, and emissions density zoning.[21]

Chapter 11

Criticisms and Alternatives

Through the 1970s there was widespread and persistent criticism of the Prevention of Significant Deterioration regulations in public comments on proposed regulations, background studies, and legal briefs. In anticipation of the 1981 amendments to the Clean Air Act, several more substantial reviews of the regulations were produced, two of them at the request of the Congress (NCAQ 1981; NAS 1981).

Most of the criticism from industrial and utility sources has been directed at the ambient increments. They have been characterized as arbitrary, irrational, ineffective, and unnecessary tertiary standards. (By *tertiary standards* it is meant that the baseline concentration plus increments define ambient limits that must be protected in State Implementation Plans and defended by additional control measures and SIP revisions if threatened.) It has been argued that the PSD provisions for area reclassification, waivers, and variances were meant to provide flexibility in enforcement, so that breaches in the increments would merely be the occasion for public deliberation and decision regarding the advisability of further deterioration of air quality (NAS 1981, p. 19); but such flexibility has been rare when proposed new sources have clearly threatened to violate incremental limits, and the procedures for reclassification of areas are complex. Offending sources have usually been resited or modified to reduce their ambient impacts. The PSD regulations have operated almost as if they were tertiary standards.

The implied standards are arbitrary for two reasons. When the increments were proposed it was frankly recognized that as definitions of significant deterioration they were inherently subjective.[1] The secondary National Ambient Air Quality Standards (NAAQS) are supposed to protect against any known or anticipated adverse effects, leaving little objective basis for the increments. Also, not much was known about the relation between precursor emissions and aesthetic effects or long-range transport and transformation of pollution. The increments were not supported by the sort of periodic review of evidence and public determination that underlies the national primary and secondary

NAAQS. In addition, they are applied to baseline concentrations that vary from region to region. These concentrations range from 2 percent to 100 percent of the secondary national standard (Business Roundtable 1980, p. 25). Not even the date at which the baseline concentration is calculated is fixed; it depends on the date at which the first application for a PSD permit in the particular area is completed. Thus, the implied standards consist of arbitrary incremental limits added to arbitrary baseline concentrations.

There are several grounds on which the PSD limits can be called irrational. The first is that they do not facilitate a balancing of the benefits of maintaining the ambient limits against the costs. For example, studies have demonstrated that industrial facilities located in hilly or mountainous terrain often have exhaust plumes that are predicted to impinge on elevated ground downwind, leading to possible increment violations (ERT 1977). These predicted local violations on remote mountainsides may not indicate a likelihood of real damages, and yet may preclude siting at what on a range of considerations might be the preferred location. The ambient increments become the controlling factor in the siting decision, even though they may not protect important environmental benefits.

The spatial pattern of ambient standards implied by the PSD regulations could also be irrational. Background concentrations generally vary from region to region with population density. Therefore—since the increments are uniform for Class II areas and even smaller for pristine, usually underpopulated Class I areas—the overall limits tend to be considerably more relaxed in areas of greater population density. The potential damages, except in scenic Class I areas, are presumably closely related to the numbers of people, buildings, commercial crops, etc. at risk, and these are greater in more densely populated areas. Since the costs to a new source of preventing incremental pollution are roughly the same wherever the new source is located, a rational pattern of control would allow less additional pollution where the damages are greater (Peltzman and Tideman 1972).

Another criticism of the ambient limits refers not to their spatial pattern but to their temporal pattern. For total suspended particulates (TSP) and SO_2, both annual average and short-term increments were defined. The short-term increments have been widely criticized. Doubts have been raised that predicted short-run violations under worst-case conditions, especially in rugged terrain, given the uncertainties of modeling, are valid indicators of potential harm. The short-term increments are stricter than those implied by a statistical distribution of ambient concentrations with the mean equal to the annual average (NAS 1981, p. 66). Also, the definition of a violation as an occurrence more frequent than once a year puts heavy weight on the unreliable extreme tail of the distribution of predicted values and the particular set of meteorological values used to generate the predictions. A more sensible statistical definition of a violation would be an ambient value not to occur or be exceeded more than a certain percentage of the time. A 1, 5, or 10 percent probability value could be reconciled with current level of protection by adjusting the annual average increment, and would imply less prediction error.

The ambient limits of PSD policy have also been termed ineffective. They are only loosely related to the air-quality-related values they are supposed to protect. They do not limit the pollutants that are directly related to the problems. Increments have been determined for TSP and SO_2. However, these pollutants are not directly responsible for visibility impairment, interstate pollution, or acidification of ecological systems, the real problems that PSD policy seems to address (NAS 1981, pp. 46–48). Most of the damage is done by fine particles in the submicron range. These differ substantially in origin and fate from their larger relatives. Most fine particles arise in the atmosphere from conversion of gaseous sulfur and nitrogen emissions into sulfate and nitrate particles. The chemistry and the atmospheric transport involved lead to the presence of these fine particles at long distances from the source, often well mixed in the atmosphere rather than in coherent plumes (U.S. EPA 1980). Models that predict TSP and SO_2 concentrations from large-source emissions are not valid for fine particles. Modeling studies that predict increment violations for TSP or SO_2 almost always do so within a few kilometers of the source under unstable conditions, or because of nearby elevated terrain (Dames and Moore 1980). Such violations, or their absence, have almost no bearing on the source's contribution to long-range pollution or to visibility or acid rain problems. In fact, maximum visibility or acid deposition impacts from changing emissions rates are likely to occur at considerable distances from the source (Latimer 1979). The absence of an increment violation does not indicate adequate protection of important air-quality-related values. Further, current PSD policy does not define any ambient limits for NO_x pollution, although it has been implicated as the fastest-growing component of acid precipitation and is a major contributor to visibility impairment, regional haze, and smog problems, especially in the western United States (U.S. EPA 1979).

The ambient increments, even those for Class I areas, are inadequate to protect against visibility impairment and acid precipitation (NCAQ 1981, p. 3.5-62). In the case of visibility, the effects of a small addition of fine particles to the atmosphere are much greater where the air is clean and visibility is high than in already polluted areas. If the visual range were about 200 miles (close to the maximum), pollution by one microgram per cubic meter would reduce it by about 30 percent; if the visual range were about 20 miles (typical in populated areas), the same pollution would reduce it by only 3 percent. Therefore, in pristine Class I areas, studies suggest that large sources could have significant adverse visibility effects even if responsible for concentrations that would not violate the Class I increments for TSP or SO_2 (LASL 1980).

In the case of acidic deposition into lakes and streams in sensitive environments, the technical reason is different. Sensitive environments are those with limited buffering capacity due to thin soils and silaceous bedrock resistant to weathering. Soft-water lakes in these areas have bicarbonate as the predominant anion. When the bicarbonate is exhausted, a threshold is reached beyond which sulfate and nitrate loadings quickly reduce pH. A sustained level of acidic deposition, without any increase, can lead to increasing acidification of sensitive lakes and streams (GCA 1980). The Class II increments prevailing over most of the country permit sustained or increased regional sulfate and nitrate loadings, and are not adequate to prevent increases in acidification of sensitive aquatic systems.

Effective control over visibility impairment and acidification of ecosystems may sometimes require additional abatement by existing sources, not just limits on the growth of emissions from new sources. It has been found that much of the fine particulate haze affecting national parks in the Southwest stems from existing smelters and power plants or from distant urban pollution (NCAQ 1981, p. 3.5-62). Most of the sulfates deposited in the Adirondacks and in southern Canada arise from existing industrial sources in Ohio, Illinois, and surrounding states (Galloway and Whelpdale 1980). To the extent that these sources are in compliance with existing State Implementation Plans, they are immune from PSD regulations.

Another weakness in the effectiveness of the PSD system is the shifting date at which baseline concentrations are defined. The original proposal for PSD regulations considered 1970 as a possible baseline date, with

1972 and 1973 as alternatives.[2] The regulations implementing the 1977 Clean Air Act Amendments defined 1977 as the baseline date. After court challenge, the current regulations define it as the first date after August 7, 1977, on which a completed PSD application is submitted for the area. This shifting baseline date has already permitted a substantial amount of emissions growth and ambient deterioration that would not have been allowed had 1970 been chosen, especially in regions like the Gulf Coast where growth has been rapid (Business Roundtable 1980, p. xii). The current definition will allow continuing increases in emissions from vehicles and minor sources in rapidly growing areas without triggering PSD review of increment consumption.

From another standpoint, it has been argued that the PSD ambient limits are redundant and unnecessary. Other sections of the Clean Air Act are available to deal with problems caused by growing pollution in clean-air areas. State Implementation Plans are required to provide for the attainment of the goal of the secondary NAAQSs, which is to protect "the public welfare from any known or anticipated adverse effects." Consequently, if the secondary NAAQSs are set correctly, there should be no need to have tertiary standards above them to deal with damages to vegetation, soils, waters, or fish. Recognizing the special sensitivity of visibility and aesthetic quality in pristine areas to even small concentrations of pollution, Congress also created explicit visibility protection for Class I areas.[3] This protection is more strict than the PSD regulations; it requires that potential new sources be denied permits if they are predicted to cause any adverse visibility effects in Class I areas. The visibility regulations also provide for the retrofitting of additional emission controls on major existing sources that are found to impair visibility in Class I areas. Finally, the CAAA set up procedures to deal with long-range transport of pollution across state boundaries.[4] In view of these other parts of the air quality regulatory program, it is argued that there is no real role for PSD controls.

Another line of argument has held that the ambient limits are unnecessary because they will not be threatened. The retirement of poorly controlled industrial facilities and their replacement by cleaner plants with strict controls and the replacement of vehicles by newer models meeting strict federal emissions standards will suffice to keep emissions levels falling throughout the country, so that reasonably careful siting to avoid local buildups in emissions density would avoid the increases in local concentrations that the Class II increments prohibit. Simulations suggest that under these assumptions even optimistic projections of

Table 11.1   Projected percentage changes in emissions of criteria pollutants, by region, between 1975 and 2000.

	Particulates	SO$_2$	NO$_x$	Hydrocarbons
Northeast	+138	+49	+24	−33
N.Y. - N.J.	−5	+5	+4	−35
Mid-Atlantic	−51	−28	+1	−33
Southeast	−18	−1	+31	−25
Great Lakes	−35	−28	+17	−32
South Central	+53	+184	+82	−9
Central	−18	+29	+45	−27
Mountain	+58	+5	+100	−21
West	+18	−22	+14	−21
Northwest	+60	+85	+61	−7

Source: U.S. DOE 1979.

industrial growth and energy production can be accommodated in all industrial regions without the ambient increments becoming an effective constraint (NCAQ 1981, p. 3.5-83). These simulations are backed by a review of the experience with PSD policy from 1977 to 1980, which suggests that few proposed industrial projects have had to be resited or modified to avoid increment violation, and that almost all of those that were affected were located in rough terrain or near Class I areas (NCAQ 1981, p. 3.5-80). This argument, however, is factually deficient. Although nationwide TSP and SO$_2$ emissions are projected to decrease, the aggregates mask wide regional variations. According to projections based on macroeconomic forecasts, as shown in table 11.1, they are expected to rise substantially in the coming decades in the Northwest, the West, and the South Central region while falling elsewhere (U.S. DOE 1979). In Texas and Oklahoma, switches from gas to coal in the wake of natural-gas deregulation are expected to raise emissions substantially (U.S. DOE 1979, p. I-14). NO$_x$ emissions, for which no ambient increments have yet been defined, are expected to rise nationwide and rise rapidly in the Northwest, West, and South Central regions (U.S. DOE 1979, p. I-23). It is clear that in rapidly growing, relatively unindustrialized areas, growth of new sources will outweigh retirements of more polluting older sources, even in the relatively short run. Over half of the PSD permits that have been issued thus far have been in the South and the West, areas that fit this description (Dames and Moore 1980, p. 36). Case studies have demonstrated that in a number of growing regions, such as the Gulf Coast, a substantial fraction of the available increment has been exhausted. One study found that "in 22 of the 92 permits reviewed, cumulative consumption of either the TSP or SO_2 increment at a maximum concentration point exceeded 50 percent of the allowable increment as modeled" (Business Roundtable 1980, p. xii).

Therefore, it is unlikely that the ambient increments, at least in areas of rapid industrial expansion, would prove to be unconstraining over another decade or more.

In fact, the opposite criticism has been voiced more frequently: that the ambient limits will hamper growth and energy development, especially in the West. This criticism is based on reports of specific projects that have encountered or projected problems with PSD limits (Business Roundtable 1980, p. xiii). Careful regional studies do not find that development will, in general, be limited, although some siting decisions will probably be affected (NCAQ 1981, p. 3.5-46).

A final and important criticism of the ambient increment system refers to the existing policy of awarding rights to increment consumption to new sources on a "first-come, first-served" basis. When a few sources are capable of affecting ambient quality enough to limit the ability of other potential facilities to site in the vicinity, this policy restricts the ability of local jurisdictions to plan for economic growth. It also creates incentives for new sources to preempt and hoard rights to pollute the air, and provides no incentives for rational land use. Even if subsequent permit applicants can theoretically obtain offsetting emissions reductions from sources already present to avoid increment violations, there can be serious misuse of land if the later applicant could actually better use the site. The costs of negotiating offsets are high, and so is the probability that monopolistic considerations will affect the negotiations. Negotiations might fail, leaving the original permit holder in possession of the right to use up the limited emission rights. In general, the doctrine of "first-come, first-served" has been discredited and long abandoned as a principle in adjudicating pollution rights (Wittman 1980).

Another set of criticisms have been directed at the individual source emission limitation, the requirement that major sources and modifications employ the "best available control technology" (BACT). This limitation is defined as "an emissions limitation based on the maximum degree of reduction for each pollutant subject to regulation under the Act . . . which the permitting authority, on a case-by-case basis, taking into account energy, environmental, and economic impacts and other costs, determines achievable for such sources or modifications."[5] The procedure for decision making puts the burden of proof on the potential new source to justify the technology proposed for the project by demonstrating that any more stringent control technology would entail unacceptable energy, environmental, or economic costs. Economic costs are not defined primarily in terms of incremental costs per ton of emissions abated; the impacts of production, output, employment, and the rate of return on investment are included prominently among the considerations.[6] The dominant issue is whether the source can "afford" more stringent controls.

This performance standard is but one of many imposed under different parts of the CAAA. New sources might be required to meet New Source Performance Standards (NSPS), to employ the BACT, or to attain the Lowest Achievable Emission Rate (LAER). Existing sources might be compelled to use Best Available Retrofit Technology (BART) or Reasonably Available Control Technology (RACT), or just to meet SIP limits. All these standards are defined legislatively in very general ways and have had to be translated into precise operational limits by the EPA or by state regulatory agencies. Some limits, such as NSPS, are derived industry by industry. Others, such as BACT, have to be determined for each individual facility. In either case, they result in widely differing marginal abatement costs among sources, in terms of lifetime cost per ton of emissions removed. If standards are set uniformly for an entire category of sources, the resulting marginal costs will vary according to the scale, expected utilization rate, location, and process design of the individual facility. Physical layout and remaining service lifetime will also affect the control costs of plants retrofitted with abatement equipment. Sometimes, the disposal costs or the reutilization value of materials removed from the waste stream affects marginal abatement costs markedly. The cost variance can be wide; a factor of 4 or even 10 would not be unusual.

Emission standards like BACT, which are supposedly determined on a case-by-case basis, can also lead to a substantial variance among sources in marginal abatement costs. The regulatory decisions are made by a number of state and regional agencies operating with different information. Legislative guidelines for these decisions are vague, and stress many considerations other than efficiency (such as the effect of compliance on energy use, inflation, unemployment, industrial output and profits, and other environmental problems).

It is likely that the "affordability" criterion in BACT determinations widens the variance of incremental control costs among sources. How much abatement cost a source can "afford" depends largely on how much of the cost it can pass along to customers or suppliers. Sources that have considerable market power can "afford" more stringent controls than sources in more competitive circumstances, so their BACT levels are likely to be more stringent, implying higher marginal control costs. In fact, in a sample of 35 four-digit industries for which incremental emissions-control cost estimates were available, the simple correlation between the level of marginal costs and the share of the top four sellers in total industry sales was significant at 0.4. To the extent that industries differ in competitiveness or individual firms face different demand elasticities, the "affordability" criterion in determining BACT requirements would systematically widen differences in marginal control costs among sources.

Beyond the effects of BACT requirements on the present costs of air-pollution abatement, it is claimed that they provide mixed and possibly undesirable incentives to innovators in abatement technology. It is argued that regulated industries have no incentive to develop or demonstrate the availability and reliability of technology that would have higher removal efficiencies, if that technology would have higher lifetime costs than the operating and maintenance costs of abatement equipment already in place (ORR 1966). Since in a growing economy the prevention of deterioration in air quality requires continual improvements in abatement technology, these criticisms are important in the long run.

The BACT requirement has also been criticized because it adds to the uncertainty and cost of industrial planning (Business Roundtable 1980). Most PSD permit applicants propose to raise emissions levels by small amounts, because they plan modifications of existing plants and can reduce emissions from old equipment. Current regulations require BACT review of all emissions that would increase beyond *de minimis* amounts. Because the tonnages involved are small, the difference between standard control technology and that determined via case-by-case review means little change in emissions and a relatively large amount of administrative work. For major sources with large proposed emissions, it has been claimed that the BACT requirement introduces uncertainty about plant design until late in the planning stage, even though in reality the BACT standard has usually turned out to be identical to the NSPS requirement whenever such a requirement exists.

In general, both industry and the state regulatory agencies have objected to the complexity of the PSD review process. The inventories of existing emissions sources maintained by most state agencies are insufficiently detailed, accurate, and current to trace the consumption of increment by emissions increases from multitudinous minor stationary and vehicular sources (NCAQ 1981, p. 3.2-20). The meteorological data and models available have often been inadequate for detailed multisource modeling of ambient impacts. Estimates of growth in major point-source emissions have often been insufficiently detailed to support plans for the best uses of available increments. Skilled and experienced people to perform case-by-case BACT reviews have been scarce in state agencies. Thus, the PSD program has seemed burdensome to the states.

Industry has also objected to the complexity of the PSD review process. The requirement that air quality monitoring data be submitted for a year prior to plant operation, the detailed modeling of ambient impacts, and the negotiation of BACT standards for virtually all pollutants emitted by the proposed source have been regarded as excessively costly and time-consuming for the degree of additional environmental protection provided by the program (Business Roundtable 1980, p. v). On the average, it has taken nine or ten months from the submittal of an initial application to the final issuance of a permit under full PSD review. Detailed modeling for PSD review is estimated to cost $15,000–$50,000, and the costs of preconstruction and postconstruction monitoring might be in the range of $100,000.

Naturally, in view of these criticisms, there have been numerous recommendations for change in the PSD program. It is tempting to put such recommendations into the broader context of reform of the entire Clean Air Act, in view of the impending revisions. The rationality of the National Ambient Air Quality Standards has been attacked—partly because of disbelief in the existence of thresholds of damage, on which they were based; partly because of new evidence on exposure levels and dose-response relationships; and partly because of the insensitivity of the national standards to regional variations in population at risk and costs of abatement (Portney and Harrison 1981). Skepticism about the NAAQS also impugns the sharp distinction in regulatory policy between attainment areas and nonattainment areas, which are now subject to quite different control mechanisms and emissions limits. However, even greater skepticism must be directed at the tertiary ambient limits (implicit in current PSD policy) that erect arbitrary incremental limits atop arbitrary baseline concentrations. In view of these uncertainties, there is reason to doubt whether air-quality standards, whether primary, secondary, or tertiary, ought to be treated as absolute and inflexible. Geographically variable standards, greater reliance on economic disincentives such as emissions charges and noncompliance fees that reflect the value of pollution abatement to society, and even a complete scrapping of ambient standards in favor of engineering controls alone have been proposed as ways of dealing with this problem.

However, the proliferation of performance standards for emissions sources is also a major deficiency of the Clean Air Act. The evolution of regulatory programs has resulted in highly differential treatment of pollution sources, depending on whether they are mobile or stationary and whether (if stationary) they are new or existing, large or small, in an attainment or a nonattainment region, and in one industry or another. It has been shown repeatedly that policies that could equalize the marginal costs of abatement among sources could reduce substantially, perhaps by half, the total cost of achieving target levels of air quality (Atkinson and Lewis 1976; Mathtech 1978). At present, a factory might be subject to SIP limits, NSPS, RACT, BACT, BART, or LAER, and each might imply a different control stringency and incremental abatement cost, depending on the regulatory agency and the factory's industry. Vitiation of these distinctions by regulations that would lead all sources that contribute to a particular pollution problem to incur the same cleanup costs at the margin would be an important contribution to efficiency.

Observers, especially economists, thus wonder why this complex regulatory apparatus should not be replaced by a simple emissions fee to which all sources in an area would be subject, or an emissions market in which all sources would buy and trade emissions rights. The complications described at the start of this chapter are part of the reason. Assessing the damages of air pollution, and each source's contribution to them, is inherently difficult. In addition, the process of regulatory change and reform tends strongly to be incremental.

There have been efforts to increase the efficiency of air-quality regulation under the Clean Air Act. The "bubble" policy, which allows regulated establishments to trade off stricter emissions controls on some sources within the plant for looser controls on others in order to achieve the same or lesser emissions and ambient impacts more cheaply, was introduced and has been expanded gradually as experience has warranted. The bubble policy has now been expanded to encompass trade-offs between different plants. Offset trading, introduced as part of regulatory strategy for nonattainment areas, has been expanded by allowing sources to "bank" emissions reductions to offset later increases. Noncompliance fees, introduced as an enforcement device in the Connecticut environmental program, were incorporated into the federal program by the CAAA of 1977.

The trick, as one discussion put it, is to open the door to regulatory reforms without tearing it off its hinges. In that spirit, I shall now discuss possibilities for change in the PSD program within the overall framework of the current Clean Air Act, indicating possibilities for broader changes but not assuming a clean slate on which to rewrite the regulations. The main alternatives are discussed in two stages: First, what should be the nature of overall limits to degradation of air quality in clean-air areas? Second, how should those limits be enforced and the limited capacity to pollute allocated among potential emitters?

There have been three distinct suggestions as to the nature of the overall limits to degradation.

The first is that there should be none, that PSD policy should rely solely on performance standards for individual new sources and on the NAAQS (Business Roundtable 1980, p. vii). The argument is basically that these will be sufficient to protect air quality because projections show that expected growth in emissions would not impinge on existing ambient standards in most areas, while there exist alternative mechanisms (such as the visibility regulations) that are capable of protecting sensitive regions now classified as Class I. This argument relies on projections of future air quality under different growth scenarios that show, by and large, that Class II increments would not be seriously threatened up to 1990.

The second suggestion is that ambient increments should be retained as overall limits to degradation and extended to cover pollutants not yet subject to PSD limits (NAS 1981). The latter include the "Set II" criteria pollutants, carbon monoxide, hydrocarbons, photochemical oxidants, NO_x, and fine particles not directly covered by the TSP increment. The argument is that ambient limits provide the most direct and reliable way to monitor changes in air quality and to fulfill the purposes of the PSD section of the Clean Air Act. Modeling, for all its weaknesses, remains the best way to predict the effects of increased emissions on air quality. In this view, further efforts are needed to extend modeling capacity to long-range transport and the atmospheric chemistry of re-active pollutants, as well as to visibility effects. Ambient increments need not be treated as inflexible tertiary standards, but can be used as triggers to initiate more detailed study of a potential source's effects on air-quality-related values.

The third suggestion is that overall limits to degradation should be expressed as area-wide emissions ceilings, or maximum emissions densities (de Nevers 1979). These could be differentiated geographically to express varying degrees of environmental protection. Area-wide emissions limits would reduce the role of modeling in PSD planning. The important damages PSD policy seeks to prevent, long-range visibility impairment and acidic precipitation, are regional problems that depend more on total atmospheric loadings of pollutants than on local concentrations. Controls on area-wide emissions densities can deal simply and effectively with these problems, whereas ambient limits create only local siting constraints in the vicinity of high ground and Class I areas and do not effectively regulate total emissions loadings. Area-wide emissions limits present opportunities for local jurisdictions to consider and decide on priorities for growth and environmmental protection. Like an area-wide "bubble" policy, they provide flexibility and facilitate tradeoffs among competing uses of the environment.

Choice among these alternatives obviously involves complex judgments. All three have points in their favor and weaknesses. Reliance on individual-source controls alone, without overall limits to degradation other than the NAAQS, most simplifies PSD policy. On the other hand, it cannot provide adequate safeguards against air-quality degradation that is occurring and expected to occur in some rapidly growing regions. As time goes on, it will be impossible to offset added emissions resulting from economic growth with reductions from retirement or retrofitting of old plants, especially in the newly industrializing parts of the country. Sooner or later, well-controlled new sources will predominate; at that point, total emissions will be determined by the growth of new sources, not by the replacement of older, dirty facilities.

Retention of ambient increments as overall limits to degradation provides the most continuity with current policy, and the most information (through its emphasis on ambient impact modeling and monitoring). On the other hand, it is the most costly and cumbersome of the three approaches. Also, a strong case has been made that the ambient increments are not good indicators of potential damages. They signal damages when none are imminent, as when a short-run increment is violated on some remote mountainside or when a Class II increment is violated at ambient concentrations well beneath the secondary NAAQS. They also fail to signal damages when they are imminent, as when a proposed facility would affect visibility in a Class I area or contribute to regional haze in such an area without violating a Class I increment, or when a facility would add to total atmospheric loadings of pollutants that would contribute to acid deposition in sensitive areas.

The adoption of ambient limits to degradation also greatly complicates the *ex ante* allocation and the *ex post* enforcement of rights to pollute by emissions sources. Each source has its individual pattern of ambient impacts, depending on its emissions characteristics and the surrounding terrain and meteorology. Prediction of these patterns is subject to all the inaccuracies and uncertainties described earlier. Thus, all sources in a locality might operate within the requirements of their emissions permits (designed to preserve ambient limits), and yet increment violations might occur. In such circumstances, it would be very difficult to know to what source to ascribe the violation or what permit condition to revise. Aside from this enforcement problem, a unit of emissions from any source would typically differ in impact at all receptor locations from a unit of emissions from another. This lack of equivalence implies that efforts to maintain ambient standards at least cost could not rely on mechanisms that minimize the cost of limiting total emissions. The degree of control required of each source would vary according to its contribution to pollution at local "hot spots" that tend to exceed the ambient limits. This implies that an emission-charge scheme designed to maintain ambient limits would require a different charge for each source that would change whenever a new source located nearby. Emissions rights that could be exchanged or traded would have unequal values per unit of emissions between any two sources. Markets for rights to pollute would have to be established not for units of emissions, but for rights to degrade ambient quality at each monitoring location threatened with increment violation. If maintenance of a certain ambient quality at each location is indeed the objective of an air-quality program, then it can be demonstrated that such mechanisms lead to more efficient patterns of abatement by multiple sources (Atkinson and Lewis 1976). In the case of PSD policy, however, it is evident that the implied ambient limits are arbitrary, and their defense is not of itself a valid objective of policy. Therefore, the added complications of allocation and enforcement of emission rights are not readily justified.

A substantial case can be made for the use of area-wide emissions limits as the overall limits to degradation in PSD policy. Area-wide emissions limits (often called "emissions density zoning") has been studied extensively (U.S. EPA 1977) and has been the subject of experimentation in limited areas.[7] It is consistent with—indeed, it is an extension of—the multiplant bubble policy initiated by the EPA. Moreover, even a system of ambient limits must be translated by modeling into emissions limits before it can be enforced, because permit conditions on the individual source are invariably written as emissions limits. When the model appropriately assumes complete mixing of the pollutant in the affected region, so that one source's emissions are approximately equivalent to another's, an area-wide emission limitation is the result. In this study, for purposes of analyzing the alternatives for allocation and enforcement of emission rights, this form of overall limit is assumed.

One option for enforcing and allocating emission limits—the base case—is to preserve the existing system. Since the inception of PSD policy in 1974, the EPA has issued construction permits to new-source applicants on a "first-come, first-served" basis. States that have taken over the PSD regulations have adopted this rule, sometimes with the proviso that no new source may reduce air quality by more than a percentage (typically 50 or 75 percent) of the remaining increment. If the potential new source's predicted emissions exceed its limit after providing BACT, it may locate elsewhere, or reduce predicted emissions by even stricter controls, or acquire offsetting reductions in emissions from other sources in the area so that the net increase will not exceed the limit.

The "first-come, first-served" rule has severe deficiencies. It provides no incentives for cost-effective patterns of emission control. It provides no mechanisms for the rational use of land. It encourages preemptive PSD applications and the hoarding of entitlements to emit. It provides asymmetrical incentives to new and existing sources. It provides no information to regulators on the marginal costs of maintaining the overall limits to degradation. It provides no flexibility in enforcing the overall limit, whatever the cost of maintaining it might be. The supply conditions facing the potential new source implied by this rule are depicted in figure 11.1. Up to the limit Q set by the percentage of the remaining entitlements allowed the individual source, entitlements are free. At that point, there is a sharp discontinuity: Further entitlements are available only through negotiations with other sources. Other sources will be willing to part with entitlements, if at all, at a price no less than

Limited Giveaway

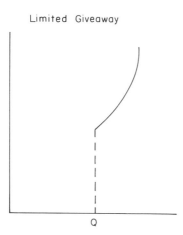

Figure 11.1 Graph illustrating limited giveaway. Potential new sources are eligible for permits to emit up to some limit Q. Emissions beyond that limit are to be balanced by offsetting reductions in emissions from existing sources.

their opportunity costs, and that will depend upon the specific market situation. At a minimum, the supply price will be the incremental costs to the source of reducing, eliminating, or refraining from the emissions-creating activity, or the costs of abating emissions caused by the activity. A rising supply schedule would be expected: The greater the volume of emissions for which entitlements are sought in a particular area, the greater the unit price. In this option, the limit Q is open to policy choice. If the percentage of the available entitlements awarded to each successive applicant is reduced, then the limit Q shifts to the left. As each successive applicant is accommodated, the resulting emissions both use up the amounts available and add to the supply of potential offsets. Thus, for the next applicant, the limit Q lies to the left, and from that limit the supply curve of entitlements from existing sources is displaced to the right.

A second option replaces the award of entitlements through the "first-come, first-served" system by the use of a market mechanism: the auction sale of emission permits up to the available limit. It would be consistent with current PSD policy to limit the bidding to potential new-source applicants, but there is no reason why the market need be restricted that way. Existing sources, currently "grandfathered," could be required to purchase entitlements for their future emissions. Environmentalists could be allowed to purchase emission rights and then not use them.

The principal advantage of the auction over the current system is that the scarcity value of clean air would be communicated to all claimants as an economic cost of pollution. This would encourage abatement and promote patterns of land use that would take air quality as well as other scarce values into consideration. An auction system that included existing sources in the bidding pool would provide symmetrical incentives to new and old claimants. What the auction mechanism and the current system have in common, however, is that they both treat the overall air quality or emission limitation as inflexible. The overall supply of entitlements is inflexible, whatever the cost of abatement to stay within the limits.

There are many possible auction types. For an allocation system based on standardized emissions in a single area, the option explored is a sealed-bid auction, in which as many bids as possible are filled at the highest unsuccessful bid price. That is, bid submissions would include the desired emissions rate, together with the information needed regarding emissions and source characteristics to calculate a standardized emissions rate, and a bid price. Bidders might, if they chose, submit bids for alternative numbers of entitlements. Starting with the bid implying the highest price per unit of standardized emissions, all bids would be accepted until the emissions limit became exhausted. These bids would be filled, not at the bid price, but at the unit price implied by the highest bid that cannot be filled. Barring collusion, this would provide all bidders with the incentive to bid their reservation prices (the maximum amounts the entitlements are truly worth to them), because a bid that is higher than necessary to win implies no excess costs to the bidder (Vickrey 1976). In theory, this system is compatible with an efficient allocation of emissions rights. A potential new source would retain the option of acquiring entitlements from existing sources through negotiation, instead of in the auction. The supply price of entitlements from existing sources would then set a ceiling on the prices bid at auction. In light of the intermittent demand for potential new sources for emissions rights in particular locations, the auction system might operate in a way similar to the offshore oil tract leasing system (U.S. DOI, n.d.).

Under an auction system restricted to potential new sources for the allocation of emissions rights, with the option of procuring offsets from existing sources, the supply conditions facing new sources are depicted in figure 11.2. (Competitive demand conditions are assumed.) The price of the amount of Q of entitlements offered for sale is determined by

Auction Market

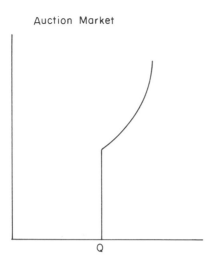

Figure 11.2 Graph illustrating auction market. The government will allocate through auction permits for emissions up to the limit Q. Potential new sources may alternatively acquire permits from existing sources by negotiation of offsets.

the demand, but constrained by the price at which entitlements would be available from existing sources. Entitlements in amounts greater than Q would only be available at a supply price determined by the cost of obtaining offsets from existing sources.

A third option is the generalization of the offset system to cover the supply of all entitlements to new sources in PSD areas. Existing policy encourages the procurement of offsets only when the limit of "free" entitlements is exhausted. Under this third system, there would be no entitlements free to the first comer. Instead, all potential new sources would be required to procure offsetting reductions from existing sources to cover a fraction of their remaining emissions after adoption of best available control technology. The fraction, or offset ratio, would be a policy variable. At present, in nonattainment areas, the determination of this ratio is the responsibility of the local regulatory agency, subject to the requirement that it be greater than one for one and high enough to ensure reasonable progress toward attainment of ambient-air-quality standards. Presumably, in PSD areas the offset ratio would also be determined by the regulatory agency, but would have to be less than one for one to ensure only "reasonable further deterioration" of air quality. The range of offset ratios would be from 0 to 1. Should there be no competition for emissions rights in a particular area and no

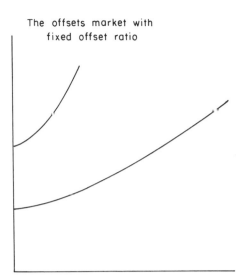

The offsets market with
fixed offset ratio

Figure 11.3 Graph illustrating offsets market with fixed offset ratio. The potential new source must surrender offsets acquired from existing sources for a specific fraction of its permissible emissions. In this case, it is as though, for each permit it acquires from an existing source, it is awarded $(1/r - 1)$ free permits by the government. The effective price of permits to the new source is r times the price of acquired permits, and the total amount of permits available is $1/r$ times the amount acquired from existing sources.

indication that air quality would be likely to deteriorate in the foreseeable future as a result of emissions from any source, then, in effect, the legal increment would be a nonbinding constraint, and the appropriate offset ratio would be near 0. At the other extreme, if it appeared that foreseeable increases in emissions, even under compliance with existing and likely future regulatory policy, would lead to violation of an ambient standard (either an increment or a secondary NAAQS), or if an increment were already found to be violated, then the appropriate offset ratio would be near 1.

In principle, the supply conditions for the individual source implied by this offset policy are identical to those under a bonus scheme: For every unit of entitlement procured from existing sources, the potential new source obtains a certain number free from the government. These supply conditions are illustrated for a 50 percent offset ratio in figure 11.3. For a 50 percent offset ratio, the new source gets a free entitlement for each one purchased. The effective supply price for the purchase of x units is $p(x)/2$, where $p(x)$ is the offset purchase price; the quantity

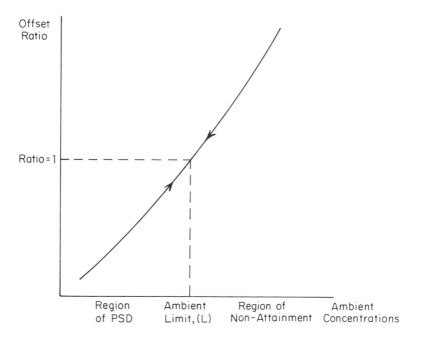

Figure 11.4 Illustration of the symmetry of the offset system in PSD and nonattainment areas.

associated with this effective price is $2x$. The effective supply schedule is a transformation of the supply schedule for purchased offsets available to the new source. If the supply price for offsets is $p = p(x)$ and the offset ratio is 4, then the effective supply schedule is $rp - p(x/r)$. As r varies from nearly 1 to nearly 0, the supply conditions change from being an approximation of existing policy, with no "free" entitlement, to an approximation of existing policy with a very large free entitlement.

The offset ratio can be regarded as a policy variable that approaches 1 as the overall emission or ambient limit is approached from either direction, as illustrated in figure 11.4. In other respects, this allocation system would operate under rules like those in effect in nonattainment areas. Offsetting reductions provided by existing sources would have to be in excess of those required by current regulations, be recorded in the State Implementation Plan, and be enforceable through compliance procedures. The regulatory agency would define the area for each substance around the site of the prospective new source within which offsetting reductions might be obtained.

This policy alternative would avoid a range of boundary problems. Areas now classified as nonattainment may become clean-air areas, either because of a real improvement in air quality or because of reclassification after an intensification of monitoring. If reclassification could invalidate decisions based on one set of regulations, enormous uncertainty would result. For example, if "banked" emissions in a nonattainment area should become worthless, if the area were reclassified as in attainment, because PSD policy disallowed offset trading, long-term planning by sources would be discouraged. Clearly, the more consistent and unified the regulatory approach in the two kinds of areas, the less complex the set of regulations facing private sources, the lower the regulatory burden, and the fewer the possible anomalies.

The creation of tradeable property rights in emissions promotes cost-effectiveness: Sources that are costly to control can be abated less, and sources easy to control can be abated more. It provides roughly symmetrical incentives for potential buyers and sellers, information on the private costs of maintaining air-quality standards, and incentives for technological innovation.

In contrast to both the "first-come, first-served" and auction mechanisms, the use of offset trading in PSD policy with an offset ratio less than 1 can provide some flexibility in the overall supply of entitlements. If the offset trading ratio is low and the supply of entitlements from existing sources is somewhat responsive to price, the aggregate availability of emission entitlements will increase with the demand for them and with the costs of controlling emissions by new sources. Provided that trading is not on a ton-for-ton basis, the market mechanism implies that overall standards are not absolute but can be exceeded if the demand for entitlements is sufficiently strong and the supply from existing sources sufficiently elastic.

The offset mechanism also contrasts with the current policy and the auction alternative in the initial assignment of property rights to use (or abuse) the atmosphere. Current policy implies that new and existing sources have a limited right to emit. They must comply with costly conditions set by the regulatory authority, like adoption of BACT controls. These costs can be regarded as a minimum charge for the right to dispose of wastes in the atmosphere. The auction policy vests these rights initially in the government, which assigns them to the highest bidder. Auctions from which existing sources are exempt represent a cross between these two mechanisms. In contrast, the offset policy acknowledges the rights of existing sources to pollute and requires new sources to acquire entitlements from them. Existing sources are granted a joint monopoly over pollution rights.

This arrangement could create problems of equity and efficiency in PSD areas. If negotiating costs were low, information readily available, and potential traders numerous, it would make little difference from the standpoint of efficiency to whom rights were awarded initially. They would end up in the hands of those to whom they were most valuable (namely, those to whom further abatement would be most costly). However, these conditions do not generally exist in PSD areas. Almost three-fourths of PSD permits so far have been issued in the South and the West, where emission densities are relatively low and the potential supply of offsets is scanty. Markets for emission rights would be intermittent, thin, and uncompetitive. Potential trades between existing and new sources might be deterred by rivalry. Control over emission rights by existing sources might provide them with a powerful barrier to entry. For example, development of a mineral lease or a synthetic-fuel project might be blocked by a neighboring energy-development company that could be the only potential source of emission offsets. The results could be highly inefficient.

The negotiating costs of offset trades between firms is evidently high. Over 90 percent of offset cases in nonattainment areas through the end of 1980 were intrafirm offsets (Dames and Moore 1980). Therefore, it is important that the initial allocation of rights is to those to whom the rights would be most valuable.

If the offset mechanism were the only allocating device, industrial development in much of the United States could be determined mainly by the availability of offsets. For example, of the 2,412 counties for which data were available, 860 had no major source of SO_2 emissions and only 657 had more than one. In a review of PSD applications through the end of 1980 it was found that most of the applicants were the only facility to emit an increment-consuming pollutant within a 50-kilometer radius (Dames and Moore 1980, p. 4-20). There is a danger that land use could be as inflexibly and arbitrarily determined by the availabiltiy of offsets as critics argue it now can be by the inviolability of ambient increments.

There is also the important issue of distributional equity. The offset system would award to existing polluting firms in PSD areas a very valuable right. There is likely to be excess demand for offsets in many PSD areas and excess supply in many nonattainment areas. In some PSD areas the existing inventory of potential offsets is small, and growth is rapid; in nonattainment areas, which tend to be older industrial areas, the inventory is often large and growth slow. Also, much of the growth in PSD areas will be from industries that are tied to specific locations or subject to multiple siting constraints. The demand for offsets from such industries is likely to be very inelastic, because there is nowhere else they can readily go. Consequently, holders of offset rights could extract substantial rents from potential entrants. Of all PSD permits issued through 1980, 42 percent were in primary energy production and distribution (excluding industrial boilers, incinerators, and the like) and over half in either energy or other primary industries like mining and primary metals and minerals (Dames and Moore 1980, p. 3-6). An offset mechanism alone would give existing sources substantial control over resource development in PSD areas. Finally, the offset system to prevent significant deterioration implies offset ratios less than 1, which means that the value of emission rights (per ton) to the seller is greater than the cost to the buyer. If the offset ratio were 0.5, the value to the seller would be twice that to the buyer. This strengthens the ability of sellers in noncompetitive situations to extract rents from potential entrants.

A fourth option, which does not have these limitations, is one that provides for government supply of additional entitlements at a ceiling price. There are several ways this might be done. A public agency might supply entitlements to any applicant at a fixed permit fee. Alternatively, new sources might be assessed a penalty proportional to their emissions in excess of those for which they hold entitlements. This is analogous to noncompliance fees, which were introduced in the Connecticut enforcement program and adopted in the Clean Air Act Amendments of 1977,[8] but were adapted to new-source review procedures rather than to enforcement of existing-source regulations.

The supply conditions that will face potential new sources whatever the procedure are depicted in figure 11.5, where the offset ratio is again assumed to be 50 percent and the price at which additional entitlements are provided or the penalty per unit of emissions assessed for violation of permit conditions is p. If p were sufficiently low relative to the offset supply price, there would be no demand for offsets and this option would be equivalent to an emissions-charge scheme. Theoretically, the price might be set at the level thought necessary to ensure protection of overall limits. This would be the estimated marginal cost of controlling emissions enough to avoid a violation. Practically, it would be impossible to use the charge in this way under new-source review procedures,

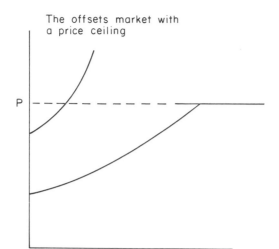

Figure 11.5 Graph illustrating offsets market with price ceiling. The potential new source must acquire offsets from existing sources for *x* percent of its allowable emissions, but the government will impose a noncompliance charge of *P* dollars per ton of emission per period in excess of permitted amount.

because the required information for charge setting could not be unobtained by the regulatory agency. The use of a charge to protect overall limits still requires that the regulatory agency know the abatement cost schedules of potential future sources, the number, and the emissions potential of each. In general, the use of emissions charges to protect standards is a more workable strategy when the problem is to control existing sources than it is when the problem is to regulate future sources. Under new-source review procedures, devices such as auctions or the creation of private markets in entitlements, in which the regulatory agency takes a passive role in price determination, are more appropriate to the actual distribution of information. Charges then establish upper limits on prices.

An alternative approach would be to set the charge level with reference to the marginal damages predicted to result from violation of an ambient limit. If the damage function could be assumed to be approximately linear, so that marginal damages were approximately the same at all ambient concentrations, then estimation of this charge level might be feasible. Otherwise, with a highly nonlinear damage function, it might be quite difficult to estimate the appropriate charge level. Whichever way the price is determined, this option differs from the preceding in a fundamental way. It implies that there is some marginal abatement cost above which it is not worthwhile from the standpoint of public policy to defend the ambient limit. Above that implied cost, the supply of entitlement becomes elastic.

These four alternatives span a wide range of control strategies for new sources in PSD areas, incorporating economic incentives and disincentives. Supply conditions for entitlements range from inelastic to elastic. The regulatory agency appears as an active charge setter, a passive supplier of entitlements, or an overseer of private market transactions. The initial assignment of rights is with existing polluters in some alternatives, but is retained by the community in others.

Chapter 12

The Performance of Alternative Systems

Each of the approaches discussed above would result in a quite different response by new and existing sources. Of primary concern to environmentalists is the degree to which air quality in clean-air areas would be protected. This would depend on the amount of abatement by new and existing sources and the net increase in emissions resulting from economic growth. Two of the stated purposes of the PSD section of the Clean Air Act Amendments of 1977 are "to preserve, protect, and enhance air quality in special natural areas" and "to reduce conflict between economic growth and air quality goals." To economists, the latter means finding an efficient level of emission control, one that balances the economic costs with the environmental benefits. Current PSD policy falls short of this objective, for reasons already explained. A prime objective of any revised PSD system would be to achieve a better balance of marginal costs and benefits.

However, because of the difficulties of identifying and estimating the benefits of the PSD program, which consist more of aesthetic and eco-logic preservation than of health protection, consideration of this broader efficiency objective is deferred to a later section. The more limited efficiency objective of cost-effectiveness—achieving whatever air quality is preserved at minimal economic cost—is examined first. Efficiency gains in this sense can be measured by the reduction in the variance in marginal abatement costs among sources in a region, with suitable weighting to account for the different emission tonnages of different sources. The only policy that could eliminate these cost differences entirely would be one that eliminated SIP limits, BACT requirements, and other source-specific emissions limitations while subjecting all sources in a region to the same emission penalty rate. (Then all sources presumably would control emissions until the marginal costs of control equaled the penalty rate.) Subjecting all sources in a region to auction bidding would not necessarily accomplish the same result, because of informational and market-structure problems. Offset trading can also be distorted by market structure, and at trading ratios less than 1 a systematic difference between incentives to buyers and sellers is present.

Rather than to assume that current policies will be scrapped altogether in favor of an emission charge, it is sensible to inquire which of the available alternatives would contribute most to efficiency if used in conjunction with elements of the present system, such as source performance standards. The total variance in marginal costs among new and existing sources can be decomposed into three parts:

$$S = bS_n + (1 - b)S_e + 2b(1 - b)(\bar{x}_n - \bar{x}_e)^2$$

where S is the weighted variance of costs among all sources, S_n is the same statistic for new sources, and S_e is the variance for existing sources. The parameter b represents the expected share of new-source emissions in total emission tonnages in the region, and the last term $(\bar{x}_n - \bar{x}_e)$ is the average difference between marginal control costs for new and existing sources. In other words, the potential efficiency gain in any region can be assessed by asking three questions:

How large is the variation in marginal control costs among existing sources, and among potential new sources?

How large is the expected growth of emissions over the planning horizon, relative to the present level?

On average, is the marginal cost of abatement higher for new or existing plants? (For example, if projected growth were slow, the variation in control costs higher for existing plants, and their control costs at the margin higher than for new plants, efficiency would be best served by a policy that concentrated on promoting cost equalization and higher incentives for abatement in existing plants.)

A final consideration is the distributional impact of the various alternatives. Regulated sources and politicians tend to be more concerned about the absolute and relative costs of regulatory measures to them and their constituents than they are about efficiency. Much of the controversy over the PSD program and the Clean Air Act as a whole have stemmed from their regional impacts (Ackerman and Haskell 1980; Navarro 1981). For PSD, these depend largely on the total locational costs of siting and operating attainment areas, which include both abatement costs and the payments the source must make to obtain emission entitlements. The higher the costs, the more likely the potential new source would be to locate elsewhere, to reduce the bid price for the site, or to regard the project as uneconomical.

Also relevant is the distribution of receipts from new sources between existing sources and the government. Government receipts might be an embarrassment to a regulatory agency, and might thus be returned in some form, or used to avert environmental damages or to promote industrial expansion. Transfer payments from new to existing sources, by contrast, might discourage entry by new sources into product or factor markets and encourage expansion by existing sources. Existing firms could represent a different industrial mix, with different employment and environmental demands, than potential new sources.

Information useful in assessing the results of alternative policies can be organized in the following categories:

the price, or effective unit costs, of emission entitlements to new and existing sources,

the amount of additional abatement (beyond SIP or BACT requirements) carried out by existing and new sources, and the net increase in emissions, and

total payments for entitlements by new sources to existing sources and to the government, and the total siting costs to new sources.

The general supply conditions for entitlements implied by each of the possible allocation systems provide part of the set of factors that would determine these results. Another part, still to be discussed, is the demand for entitlements on the part of potential new sources. Demand might be strong or weak relative to the emission limits set by the regulatory agency or the emission entitlements potentially available from existing sources. In addition, it should make a difference (particularly when market processes such as auction bidding and offsets transactions are part of the allocation process) whether the demand for entitlements arises from a single potential new source or from many; that is, whether the market for entitlements is competitive or monopsonistic. Correspondingly, it should be relevant whether the potential supply of entitlements from existing sources stems from a single monopolistic seller or from many potential sellers. It cannot be assumed that there will be many potential buyers or sellers of entitlements in PSD reas.

For analysis of market behavior, the possible conditions in the market for entitlements are typified by three limiting situations: competition in both demand and supply, monopsony (in which a single new source wishing to locate in the PSD area would face numerous potential suppliers of entitlements), and monopoly (in which there would be a single existing source potentially willing to supply entitlements, and many

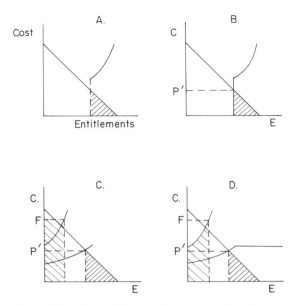

Figure 12.1 Competitive market structure: low demand.

new sources as potential buyers). These extreme cases are sufficient to establish the range of performance likely to occur, and give rise to determinate behavioral relationships under assumptions of profit maximization whereas intermediate situations of oligopoly or bilateral monopoly may not.

Performance

As an aid to understanding each policy option, the equilibrium positions under each one for the same set of demand conditions are displayed for the competitive case in figures 12.1 and 12.2. The supply conditions implied by each policy option are already familiar. The demand conditions are approximated by linear functions. The maximum price a potential new source would be willing to pay for a one-unit entitlement would be the incremental cost of abating that unit of emissions. Demand from a potential new source in an area would fall to zero when the total siting costs, including both abatement costs and payments for entitlements, exceeded the differential site rent that could be commanded by that location. The demand function represents the combined demands of all potential new sources in the area, although this might be a single firm.

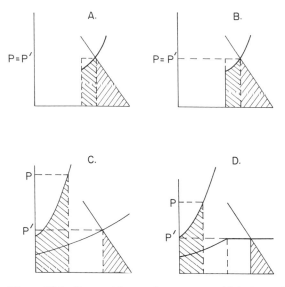

Figure 12.2 Competitive market structure: high demand.

The simplest conditions, represented in figure 12.1, are those of competitive demand and supply, when the market demand for entitlements is low relative to the available supply. They still illustrate the differences between alternative policies. Under the present policy of "limited giveaway," some sources are still induced to control emissions beyond BACT requirements in order to site. This is displayed in figure 12.1A as the distance between the zero price point in the demand function for entitlements (the emissions that would result if there were no charge for emissions after BACT controls) and the vertical limit in the supply curve (the emissions for which entitlements are available from the regulatory agency). The additional costs (beyond those for BACT controls) incurred by potential new sources for abatement are represented by the shaded region under the demand curve for entitlements, which is the sum of marginal control costs beyond BACT levels for new sources. Existing sources undertake no additional controls, however, because their minimal supply price for offsets is above the marginal control costs of new sources for all emissions not covered by entitlements provided free by the regulatory agency. Therefore, there are no payments for entitlements by new sources, and the total locational costs are simply the costs of additional abatement. Emissions increase to the limit set by the regulatory agency. There are no stimuli toward equalization of marginal costs for new or existing sources.

Under the policy of allocating entitlements through auction sales, depicted in figure 12.1B the main difference is that entitlements are likely to be captured by the potential new sources to which the locational advantages of the site are greatest. Bid schedules specifying the highest unit prices at which entitlement will be purchased will be honored until the supply is exhausted, and filled at a price p', which represents the highest unsuccessful bid. Total payment for auction purchases of entitlements is represented by the rectangle bounded by the price p' and the supply limit. As in the policy of limited giveaway, some new sources are induced to apply additional emissions controls beyond BACT, and at the margin the price paid for entitlements (p') is approximately equal to the marginal cost of abatement. Unlike the giveaway option, this policy promotes a cost-effective pattern of abatement among new sources. Net emissions increase under these two policy options by the same amount; only the allocation of entitlements differs.

Figures 12.1C and 12.1D represent the results of allocation through the offset market, with a 50 percent offset. Because p' (the effective price paid by new sources under these assumed low demand conditions) is lower than the price at which entitlements are available from the regulatory agency, the two alternatives lead to the same performance. However, they contrast sharply with alternatives A and B. Potential new sources acquire entitlements from existing sources until the effective price p' threatens to exceed the marginal costs of additional controls beyond BACT. Thus, new sources are induced to spend substantially to reduce emissions beyond BACT levels, and also to pay substantial amounts to existing sources to buy entitlements. Since each unit purchase implies a "free" unit of entitlement sanctioned by government, the price paid to existing sources is twice the level of the effective price. This price p' is the incentive to existing sources to abate beyond the requirements of the state implementation plan, and evokes a reduction in emissions from existing sources represented by the distance along the entitlements axis under the supply curve from existing sources. Both new and existing sources tend to equalize marginal control costs among themselves, but at different levels. Unlike options A and B, the offsets market provides much stronger incentives to existing sources to undertake further abatement than to new sources. Also, net emissions are lower for the same demand conditions. In that sense, offsets facilitate industrial expansion by creating inducements for the existing sources to make room for growth by reducing emissions. On the other hand, the total locational costs of new sources that do site are higher than under options A and B—a great deal higher than under option A, the limited giveaway.

The effects of changing the offset ratio depend on the elasticity of the demand for entitlements. Typically, demand is inelastic, because the marginal costs of emissions control rise steeply at control levels beyond BACT. Thus, a lower offset ratio implies lower expenditures by potential new sources for entitlements and lower prices offered to existing sources. Thus, lowering that offset ratio reduces the siting costs imposed on new sources but expands the increase in new emissions resulting from the growth of new sources.

Figure 12.2, which displays the equilibrium results of the four options under competitive market conditions with high new-source demands, demonstrates the sensitivity of the results of the policies to market conditions. In 12.1A and 12.2B it is assumed that the marginal abatement costs of new sources are higher than the supply price of entitlements from existing sources after the available entitlements from government are exhausted, so that there are purchases of offsets. Both new and existing sources equalize marginal control costs at the same level, which implies cost-effectiveness. The amount of abatement beyond BACT carried out by new sources is greater than under options C and D, although when new-source demands were low the opposite was true. In the same vein, with high demand from potential new sources the total locational costs under option B (the auction market) become higher than those under either variant of the offset-market policy.

Figures 12.2C and 12.2D show the difference (in the functioning of the offset market with high demand) that is due to the ceiling price created by official releases of entitlements. Without this ceiling, strong demand leads to very high prices paid in the offset market and much higher abatement incentives to existing sources than to new sources. Under policy D, potential new sources are induced to turn to official sources at the ceiling price. This reduces total locational costs to new firms, but it reduces incentives to abate to both new and existing firms and results in a larger net increase in emissions.

When there are sufficiently few potential new sources that their market behavior becomes monopsonistic, the same supply conditions, policy alternatives, and demand levels will typically result in quite different results in offset and auction markets. In an auction sale under monopsony conditions, entitlements would go to the sole new source either at zero price or the public agency's reservation price. In the offset market for entitlements, the new source, taking account of the impact of its demand on the price of entitlement, would be induced to purchase fewer entitlements and abate more of its own emissions. Abatement beyond BACT levels would proceed until the incremental cost equaled not the effective price of entitlements, but the effective marginal cost. Under these conditions, the market allocation system illustrated in figures 12.2C and 12.2D would lead to a high degree of abatement beyond the BACT level by the new source, some abatement by existing sources, little net increase in emissions, and relatively low payments by new sources for entitlements. This would be fairly cost-effective if abatement costs at BACT levels were lower than those facing existing sources. The new source's bargaining strength keeps the price of entitlements sufficiently low that the price ceiling remains ineffective.

These results stand in contrast to those that would be expected if the bargaining strength were on the supply side of the market, because there are few existing sources and more numerous potential new sources. In this situation, the auction mechanism would lead to performance as in competitive conditions. However, if potential new sources are forced into the market for entitlements with only existing sources as suppliers, the structure of incentives, the distribution of cleanup, and the level of total siting costs are markedly affected. The monopoly position of the existing source results in high prices for sales of entitlements and differences in incentives between the existing source and new sources. This would be cost-effective only if the existing source had much lower marginal abatement costs than new sources. Total siting costs would be high, because of extra abatement costs and substantial transfers from new to existing sources.

It is clear that the consequences of each policy option will depend on the market conditions under which it is applied. The market structure, the strength and elasticity of the demand for entitlements, the price at which entitlements are available from existing sources, and the elasticity of supply will determine the results of each policy. A numerical simulation based on the analytical models discussed above explored the four policy options under conditions of monopoly, monopsony, and competition and a variety of demand and supply conditions. The sensitivity of results to variations in the offset ratio and the ceiling price was also investigated. The main conclusion of the simulation was that the consequences of policy, with respect to emission growth and abatement, locational costs and distributional impacts, and incentives to new and existing sources, are highly dependent on the particular market conditions under which the policies would be applied. Rarely would the results of one policy in any of these dimensions be stable relative to the results of alternative policies across all the market conditions investigated.

The current "limited giveaway" policy and the auction device naturally maintain the most certain control over the growth of emissions, because of the inflexibility of the aggregate supply of entitlements. However, for given demand and supply conditions, emission growth would usually be least under the offset policy, which induces the most additional abatement by existing and new sources.

The total financial burden on new sources is usually least under the current "limited giveaway" policy, because of the free entitlements, but this does nothing to promote rational land use. The financial burden of other policies could be relatively high or low, depending on the market circumstances. The offset mechanism could be particularly burdensome to new sources if demand were strong and supply monopolistic.

The implications for cost-effectiveness are complex. The current system least promotes efficiency, because neither new nor existing sources are led to equalize marginal abatement costs; a uniform penalty rate to which all sources would adjust their abatement costs would most promote cost-effectiveness. However, under present regulations, the degree of variance among sources in abatement costs will depend on the procedures under which BACT determinations and SIP limits are made. If those source-performance standards were set in such a way that different emissions sources would incur similar costs in reducing emissions to the last (marginal) ton required by the standard, the degree of inefficiency in the current system could be much reduced. Alternatively, the scope for controlled tradeoffs in emissions reductions among existing plants under the current "bubble" and "multiplant bubble" policies (44 F.R. 71780, December 11, 1979) could be expanded and liberalized to promote cost-effective abatement patterns among existing sources.

If present SIP and BACT performance standards are maintained, then the contribution of each of the alternative policies to efficiency becomes dependent on the particular circumstances in the region, as indicated by the decomposition analysis. The relative importance, the amount of variance in incremental control costs, and the relative height of incremental control costs for new and existing sources become relevant considerations, along with the structure of market demand and supply for entitlements.

Market Conditions in PSD Areas

Demand for entitlements from major new sources

The results of alternative allocation systems will be influenced by the level and the structure of demand. The level of demand will vary from region to region and pollutant to pollutant, depending on regional growth. Projections discussed above suggest that new-source growth will be faster, relative to the existing inventory, in the "sun belt" and the Northwest, and slower in the older industrial areas. Growth is also likely to be faster for NO_x and SO_2 than for hydrocarbons and particulates. In most regions of the United States, the growth of major new-source emissions over a 10- or 15-year horizon will be small relative to the current inventory; thus, policies that would do little to promote cost-effectiveness in abatement of existing emissions would be inferior.

It cannot safely be assumed that the demand for entitlements will be competitive, because under present regulations only large new sources, with allowable emissions (after adoption of BACT) greater than 100 tons per year, are subject to all the requirements of PSD policy. The more numerous smaller sources are exempt, and large new sources may be few.

It is difficult to obtain source-by-source projections of major new emitting facilities. However, announced investment plans (including construction in progress) in industries that contribute the bulk of SO_2 and particulates emissions provide an indication. Although the data are incomplete, it appears that it will be atypical for a large number of major new sources to be siting new facilities within the same air-quality-control region in the same year. According to recent experience the most likely number of new sources actively seeking permission to site within an AQCR is one. However, in certain highly industrialized and rapidly growing regions there might be as many as twenty. This suggestion that competitive conditions in the demand for entitlements among major new sources might be atypical is borne out by a review of past applications for PSD permits. According to Dames and Moore 1980, from the passage of the Clean Air Act Amendments of 1977 to early 1980 there were 1,092 permit applications for potential new sources or modifications of existing sources. Their distribution by industry and region is given in table 12.1. The geographical pattern is informative. Only 25 percent of the applications were from regions I–III and V— the older North Central and Northeastern industrial regions. Over 60 percent came from regions IV, VI, and IX—the South and the South-

Table 12.1 Numbers of PSD permits issued in various industry groups and EPA regions.

	EPA region										Total
	I	II	III	IV	V	VI	VII	VIII	IX	X	
Electric generators	0	7	24	19	14	29	17	4	16	4	134
Industrial boilers	3	6	8	13	7	3	1	1	5	2	49
Steel	0	5	8	8	8	11	2	1	2	0	45
Aluminum processing	0	0	1	7	0	0	0	0	0	5	13
Lead processing	0	0	2	1	0	1	2	0	0	1	7
Copper processing	0	0	0	1	0	1	0	0	0	0	2
Other mineral/metal processing	1	0	0	11	1	5	0	1	1	0	20
Petroleum refining	0	7	2	8	0	61	5	3	15	3	104
Other petroleum products	3	13	20	68	3	23	16	2	6	0	154
Chemicals	0	19	6	15	4	31	4	0	5	0	84
Paper mills	1	1	1	15	0	8	0	0	1	4	31
All other	15	38	29	142	24	79	20	28	65	9	419
Total	23	96	101	308	61	252	67	40	116	28	1,092

Source: Dames and Moore 1980.

Table 12.2 Existing sources of particulates: distribution of counties by number of plants and number of SIC codes.

No. of codes	No. of counties with specified no. of	
	Plants	SIC codes
0	252	252
1	666	816
2	435	471
3–5	596	574
6–10	288	210
>10	175	99
Total	2,412	2,412

west. In the older industrial regions, PSD applications came at approximately the rate of one per AQCR per year over the period. In the Southern regions this rate was closer to three per AQCR per year, but even that higher rate is not sufficient to ensure competitive conditions in the demand for PSD permits from potential applicants.

The condition of supply of entitlements from existing sources
For the same reasons that suggest that demand conditions for entitlements might be uncompetitive in many PSD areas, it is likely that there would be few potential sellers of entitlements among existing sources. There could be few large emission sources in an area, and among the sources that do exist, some might be unwilling to provide offsets, either because the offsets were reserved for future internal expansion or because withholding potential offsets would create a barrier to the entry of possible competitors.

Emissions inventory data have been tabulated at the county level for sources with actual emissions greater than 50 tons per year and the number of different four-digit Standard Industrial Classification codes represented by those large sources (tables 12.2 and 12.3). In most counties there are no more than two large potential sources of entitlements. The total number of large sources indicates the degree of monopoly inherent in the market structure only approximately, since it ignores smaller sellers and equates the county with the likely market area for exchange of entitlements. The total number of different SIC codes represented among large sources in each county provides an estimate of the probability that any given potential new source would have to negotiate for entitlements with an existing source in its own

Table 12.3 Existing sources of SO$_2$: distribution of counties by number of plants and number of SIC codes.

| | No. of counties with specified no. of | |
No. of codes	Plants	SIC codes
0	860	860
1	657	703
2	312	318
3–5	347	340
6–10	132	130
> 10[a]	102	59
Total	2,412	2,412

a. Source: unpublished EPA data.

industry. The greater the number of SIC codes present, the greater the likelihood that any potential new source could find sellers of entitlements from other industrial sectors. Negotiations among buyers and sellers of entitlements within the same industry are complicated by issues arising from oligopolistic behavior in product markets.

There are substantial differences in marginal control costs among existing emissions sources. Public information on this variance is inevitably an underestimate, because data always represent some degree of averaging across sources. As an illustration, table 12.4 shows the range of marginal control costs for hydrocarbons at current control requirements for a sample of sources in the New York metropolitan region, estimated as part of a study commissioned by the National Commission on Air Quality (GCA 1980). Even the range of median cost estimates by source category ranges from a high of $1,652 per ton of hydrocarbons abated to a minimum of $146 per ton saved through emission abatement. (The saving stems from the value of materials recovered.) Other studies show equally wide variations (Atkinson and Lewis 1976; Bingham et al. 1974; Mathtech 1978).

More dramatic are the potential increases in cost-effectiveness revealed by industries in response to the EPA's bubble policies. Some adjustments proposed by sources will result in 60 percent reductions in annual abatement costs, with no deterioration in air quality (Speciner 1980). A bubble policy devised for the can coating industry that allows tradeoffs in the degree of abatement on different coating lines to maintain constant total daily hydrocarbon emissions is expected to save $107 million in capital costs, $28 million per year in operating costs, and 4

Table 12.4 Cost-effectiveness of hydrocarbon controls (1980 dollars/mg HC controlled).

Source category	Range[a]		Median cost-control options[a]
Automobile coating	0 –	4,305	1,326
Can coating	123 –	1,250	516
Coil coating	21 –	166	72
Wire coating	145 –	372	229
Fabric coating	(109) –	114	44
Paper coating	(95) –	196	55
Metal furniture	563 –	3,526	1,652
Large appliances	(1,156) –	1,637	258
Degreasing	(565) –	440	(146)
Cutback asphalt	(144) –	16	(128)
Misc. refinery sources	(221) –	8	(152)
Fixed roof tanks, bulk plants, gasoline terminals	(325) –	42	(125)
Gasoline stations	253 –	718	360

Source: GCA 1980.
a. Parentheses indicate cost savings.

trillion BTU per year in natural gas consumption within firms in that one industry (EPA Office of Planning and Management 1980). These savings represent potential improvements in cost-effectiveness through equalization of marginal abatement costs within single establishments, and are only indicative of the degree of variation in marginal abatement costs among establishments.

In the absence of a conscious attempt to set emissions standards to equalize marginal control costs, one would expect the variance in costs to be larger for particulates than for SO_2 emissions and larger for existing source standards than for new source standards. A wider variance for particulates is likely, because the sources of emissions and the applicable control technology are more diverse. Fugitive and other nonstack emissions from a variety of industrial construction activities contribute a large percentage of total emissions particulates, whereas most sulfur emissions stem from either large-scale fossil-fuel combustion or mineral reduction.

Similarly, new units within any source category are likely to be similar in technology, expected lifetime, and scale. These similarities would reduce variations in emissions-control costs. New units would also escape variations in costs arising from problems of retrofitting control equipment to existing plants and sites. Therefore, other things being equal, one would expect greater cost variations among existing units than among new units. However, the EPA's bubble policy already provides a means whereby existing sources in an area can improve the cost-effectiveness of their pollution controls.

If in a region the existing and projected new-source emissions over the planning horizon were about equal, then the difference between the average control costs of new and existing sources might become an important indicator of potential efficiency gains. Should the level of control costs be higher for existing than new sources, then in the interest of efficiency a policy option that provides higher incentives for new sources should be preferred, and vice versa.

In clean-air areas, the relevant comparison is between the level of incremental costs with BACT for new sources and the level of incremental costs at SIP limits for existing sources. In general, BACT requirements are more stringent than SIP limits in terms of percentage reduction of uncontrolled emissions. Thus, with rising marginal costs, new-source cost levels should be higher. However, there are substantial additional costs associated with retrofitting control equipment to existing facilities instead of incorporating it in the design of new plants. Also, new plants are likely to be larger than older ones, and there are usually economies of scale in control technologies that involve substantial capital investments. Finally, since the investments in control equipment on existing facilities are likely to be amortized over a shorter remaining lifetime, the annualized costs might be raised considerably. Therefore, it is not unlikely that the costs of added controls on existing sources might be higher than BACT costs in many industries. Whenever this is true, policies that induce more additional cleanup by new sources (such as the auction mechanism) are favorable to cost-effectiveness.

In summary, the following generalizations about the contribution of various policies to cost-effectiveness can be made:

• The current policy of "limited giveaway" on a first-come, first-served basis contributes the least to efficiency until free entitlements are exhausted and offsets come into play. Limited giveaway does not provide incentives to cost-effectiveness to new or existing sources.

• Offset trading between new and existing sources, especially at trading ratios less than 1, cannot be counted on to promote cost-effectiveness because of the probability of thin and uncompetitive markets. Where competitive markets are possible, or if supplemented by official supply of entitlements at a ceiling price, offset trading can promote efficiency.

This approach is particularly appropriate if it is also desired to induce existing sources to undertake more of the additional cleanup, should their marginal abatement costs be lower.

• In most PSD areas, existing sources will account for most emissions over an intermediate planning horizon. Policies that promote cost-effective patterns of abatement among existing plants, like the bubble policy or a uniform emission charge, can be most productive, especially if BACT determinations for new plants are then based on a uniform incremental control cost criterion.

• In regions where emissions from new sources are expected to grow rapidly, cost-effectiveness could best be promoted by establishing a charge for entitlements, either through an auction market or through a fee mechanism. Which of these devices might be more suitable depends partly on broader efficiency considerations addressed in the next chapter.

Incentives for innovation and long-run abatement costs
Because continuing economic growth with limited natural assimilative capacity for wastes implies that emissions per unit of output must fall continually, the most important criterion by which to compare regulatory policies may be the extent to which they provide appropriate incentives for innovation in pollution technology (Schultze 1977, p. 26). Standards provide no positive incentive for polluting sources to develop technology to abate more than the standard requires, and may induce sources to obscure or delay the availability of technology to meet standards. Standards shift the informational burden onto the regulatory agency, and they discourage research that might lead only to the imposition of even tighter control requirements.

Charges, fees, and markets encourge innovation by decentralizing the search for control alternatives within appropriate price signals, and by encouraging sources to find cheaper ways to abate the remaining emissions. Markets also automatically raise incentives to innovate automatically as the pressure on the fixed environmental resource increases.

This chain of argument rests on the assumption that polluting sources are the main source of innovation in abatement technology. However, that is very often not the case. For example, electric utilities have not been the main source of innovation in the technology of electric power generation; innovation has come largely from suppliers of equipment and materials to the polluting industries (Mansfield 1977). Environmental services and control equipment are provided by a large and growing industry (Sellars and Lorenz 1979). Firms that develop new solutions to their own pollution problems often enter this industry to sell their innovative approaches to other companies.

The importance of the pollution-control industry has important implications. Vendors pursue product innovation to maximize future profits, not process innovation to reduce compliance costs as the polluting industries do. Product innovation will be sensitive to the size and the certainty of the potential market, both of which will be strongly affected by regulatory policies (Caves 1975). For example, the promulgation of a technology-forcing standard might be an occasion for foot dragging by the polluting industry, but might open a major new market for its suppliers. Their research-and-development responses might be quite different. For this reason, supplying industries have been instrumental in the development and introduction of catalytic converters for automotive emissions and scrubbers for power plant emissions (Krier and Ursin 1977, p. 158; Ackerman and Haskell 1980).

In general, standards of the BACT variety can provide strong incentives for innovation by vendors, by increasing the size and certainty of the market for control devices or improved process equipment (Repetto 1980). This was not lost on legislators who prescribed BACT standards (Stewart 1980, p. 91). On the other hand, effluent charges and similar policies might provide quite diffuse and relatively weak incentives to innovation if the polluting industries are technologically diffuse and fragmented. Which policy approach will provide the more effective incentives will depend in particular cases on the technological and market structures of both the polluting industry and the supply industries.

Chapter 13

Broader Efficiency Goals

The ultimate efficiency goal is to minimize the sum of abatement costs and environmental damage. This goes beyond ensuring that whatever cleanup is required is accomplished at minimum cost, and examines how much control is desirable. Ideally, the original damages from emissions should be equated to the marginal costs of abatement. Current Prevention of Significant Deterioration policy falls short of this goal for many reasons already discussed. A fundamental underlying problem is that the Environmental Protection Agency faces great uncertainty about damages, even more than about abatement costs. Its goal should be to use to best advantage what knowledge is available, and to elicit as much additional information and as little misinformation as possible.

In most PSD areas with ambient concentrations well below the secondary National Ambient Air Quality Standard, total marginal damages from increasing pollution are unlikely to include significant damages to health. Primary NAAQSs are set to prevent health damages with an ample margin of safety, and the secondary NAAQSs are still more stringent. Even if dose-response functions for health effects are approximately linear all the way to zero concentration, implying constant marginal damages, marginal damages in PSD areas are likely to be small because population densities are low. In counties subject to PSD regulations, average population densities were 71 persons per square mile in 1975, compared with 703 persons per square mile in nonattainment areas.

Health is not the main concern of PSD regulation. Preservation of aesthetic values and visibility are explicit among the objectives of legislation, and there is evidence that the value of aesthetic benefits can be substantial. In one study based on expressed willingness to pay for improved air quality (in southern California, where pollution concentrations are severe enough to pose significant threats to health as well as to visibility), respondents' valuation of visibility benefits was of the same order of magnitude as their valuation of health benefits, especially the reduction of chronic health effects (U.S. EPA 1979b).

By nature, the visibility damage function is quite different from that for health effects with respect to the dose-response function and to population exposure.

Take the latter first. Even though few people live in the national parks and wilderness areas, many people would be affected by any impairment of their aesthetic qualities. Large and increasing numbers of visitors travel considerable distances at substantial cost to see them, and many people who do not visit them are still willing to pay to prevent their destruction, including impairment of visibility. Therefore, it cannot be assumed that the numbers of people exposed to aesthetic losses in PSD areas is necessarily small because residential density is low.

Most man-made impairment of visibility is due to light scattering by fine particles in the 0.1–1.0 mm range, although NO_2 absorbs light and results in bluish discoloration in exhaust plumes. Fine-particle aerosols are dominated by sulfate and nitrate particles, which account for over half of actual man-made impairment. There is a close relationship among fine-particle concentrations, scattering coefficients, and visibility, whether measured by range or by effective contrasts (U.S. EPA 1979a). This relationship is one of declining marginal damage; clean air is affected much more dramatically by a given addition of fine particles than is dirty air. If the visual range is about 200 miles (close to the maximum), pollution by one microgram per cubic meter of fine particles will reduce visibility by about 30 percent; if the visual range is about 20 miles (typical of populated areas), the same pollution will reduce visibility further by only 3 percent. This relationship is plotted in figure 13.1. The valuation of visibility loss does not alter this pattern of declining marginal damage. Visibility is a public good characterized by extreme jointness of supply.

Valuation attempts have to contend with the "free rider" problem: Potential beneficiaries of visibility protection might misstate their willingness to pay for benefits because of nonexcludability. Attempts to measure benefits through sophisticated polling techniques do not indicate sharply rising marginal valuation of visibility with diminishing range. In fact, the opposite appears true. As table 13.1 shows, willingness to pay for a given increase in visual range (in this case, 25 miles in southwestern Colorado) diminishes as visual range declines. This is plausible at "scenic" ranges; the increase in visible area that results from a given improvement in visual range is proportional to the initial range, and an improvement in visual range implies an improvement in clarity at all distances.

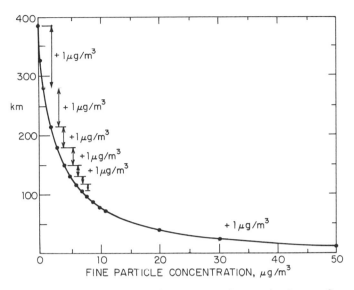

Figure 13.1   Effect of fine-particle concentration on visual range. Source: U.S. EPA 1979a, p. 2–37.

Table 13.1   Average willingness to pay (equivalent surplus) for improvements in visibility, by initial visual range.

Visibility change (miles)	Average willingness to pay	
	Residents ($/month)	Nonresidents ($/day)
50–75	4.75	3.00
25–50	3.53	2.53
25–75	6.54–7.58[a]	4.06

Source: Rowe et al. 1980, p. 10.
a. Higher figure after prompting.

Consequently, for visibility losses and aesthetic damages one is confronted with a highly nonlinear damage function. There is also substantial
evidence that, in the relevant range, the marginal-cost curve for abatement of emissions of particulates and SO_2 is quite steep. The relevant
range is the control of residual emissions after the adoption of Best
Available Control Technology. BACT, for important stationary-source
categories like industrial and utility boilers, implies removal of 98–99
percent of particulates from the exhaust stream and 90–96 percent of
SO_2. These are stringent controls. At this high level of removal, incremental costs rise rapidly. As an illustration, figure 13.2 plots the implicit
supply curve for sulfur abatement for two industrial Air Quality Control
Regions, drawn from data on actual emissions inventories in those
areas and on abatement cost schedules for each industry. It is evident
that the charges rise rapidly when emissions are reduced to 10–20
percent of uncontrolled emissions. Since these supply curves take into
account cost variations across industries as well as the possibilities for
more stringent controls within industries, they are less steep than the
curves for any individual emissions-source category would be. A similar
cost curve for sulfur emissions in southern California also shows rapidly
rising incremental costs at high levels of abatement, as shown in figure
13.3. Therefore, it can safely be concluded that at the prescribed levels
of control—BACT for potential new sources and RACT for existing
sources—the marginal costs of abatement are rising steeply.

The combination of declining marginal damages and rising marginal
costs gives rise to only two main possibilities, illustrated in figure 13.4.
At zero pollution level, the marginal damages might be so high that it
would be efficient to prevent any impairment of visibility at all; the
marginal costs of abatement rise, but the marginal benefits rise faster,
as in column a of figure 13.4. Alternatively, it might be excessively
costly to prevent impairment altogether. If this is the case, then it is
possible that the marginal damages fall faster than marginal abatement
costs, as in graph i of column b, so that no abatement to protect visibility
is warranted. However, it is also possible that the costs decline more
rapidly, as in graph ii of column b, so that it becomes efficient to halt
visibility impairment through some abatement program.

Current policy reflects the judgment that the aesthetic quality of most
Class I areas is so valuable that no impairment is justified; that the
marginal damages of any impairment exceed the cost. The confinement
of the visibility-protection program to these areas implicitly reflects the
judgment that in the rest of the country the marginal benefits are below

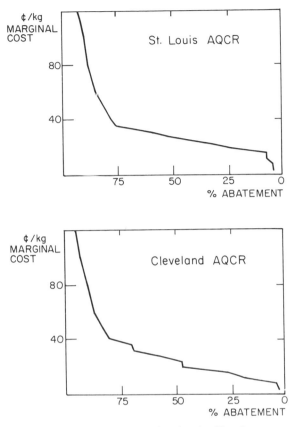

Figure 13.2 Marginal costs of regional sulfur abatement—two examples. Source: Bingham et al. 1974, p. 105.

the marginal costs at unpolluted levels. The rationale for this dichotomous policy is strengthened by the recognition that there can be substitutions among pristine and polluted regions, both in the location of industrial facilities and in the enjoyment of recreational resources. The valuation of visibility in one region depends partly on the availability of unspoiled vistas elsewhere. Similarly, the value of industrial sites in one area depends on the availability of sites in other places. The problem is really one of deciding where visibility should be protected and where polluting production facilities should be located.

The essence of the broader problem can be captured by an analysis of two regions (Repetto 1981). Each has valuable scenic resources that are susceptible to visibility damages. Visibility in each region depends

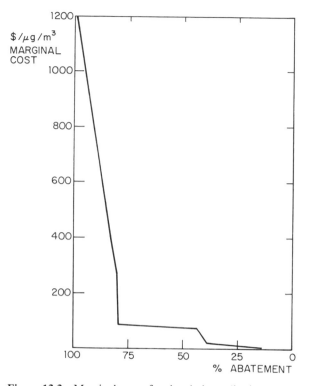

Figure 13.3 Marginal cost of regional air-quality improvement determined in California South Coast Regional Sulfur Study. Source: State of California 1978, table 8.7.

on the concentration of ambient fine particles in that region. However, the valuation of any given level of visibility in a region is influenced by the level of visibility in the other and by the level of income in the two regions. Higher income implies a higher valuation of visibility, and a lower level of visibility in one region implies a higher value of visibility in the other. Finally, given abatement technologies, the level of ambient pollution in each region depends on the level of income and output in that region. Output generated in one region substitutes perfectly for output generated in the other.

The regulatory problem is to choose the level of visibility in each region; or, equivalently, given the abatement possibilities, to choose levels of production and corresponding ambient concentrations in each region. The efficient solution is one that maximizes total welfare, including both the value of visibility in each region and the value of output. It is even more likely in the two-region case that the efficient

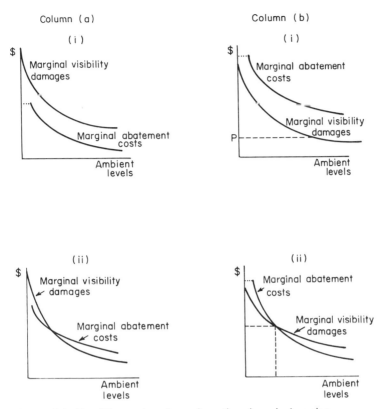

Figure 13.4   Possible cost-benefit configurations in a single region.

solution will be to concentrate output in one region and to preserve visibility unscathed in the more scenic region. This can be demonstrated by a graphical analysis analogous to that for a single region. Assume initially that abatement possibilities are the same in both regions, in the sense that the amount of income given up in reducing ambient concentrations is the same throughout. These abatement possibilites are depicted in figure 13.5 as the marginal-cost curve, which rises with successive reductions in pollution. The graph is drawn with respect to levels of pollution in region two, indexed by ambient concentrations there. Marginal visibility damages are defined as willingness to give up a dollar of income to improve visibility incrementally in region two. These are drawn as declining with increasing pollution levels, for reasons justified above. Superimposed on the graph is an additional function representing marginal willingness to pay for visibility in region one,

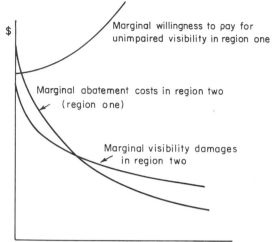

Figure 13.5 Benefits and costs of visibility protection in the two-region case with equal abatement costs in both regions.

when visibility is still unimpaired there, as a function of the level of ambient concentrations in region two. This is shown as a rising function of ambient concentrations in region two, for two reasons: first, since visibilities in the two regions are substitutes, the worse pollution becomes in region two the more good visibility is valued in region one; second, the more income is generated in region two, the greater the demand for scenic values, and the higher the valuation of unimpaired visibility in region one. It is also clear from figure 13.5 that region one is by definition the more "scenic," in the sense that at zero pollution levels in both regions there is a greater willingness to pay for visibility in region one.

To make the analysis meaningful, the marginal costs of abatement when pollution is nil in both regions exceed the marginal value of visibility. The nature of the efficient solution can be understood intuitively by considering the initial location of production. It would be the less scenic region, because the net gain in welfare (the gain in income less the loss of visibility values) would be maximized thereby. However, this dictates the location of the next production facility, because the pollution resulting from the initial production raises the marginal value of visibility in the more scenic region and lowers it in the region in which the first unit was sited. The net gain in welfare is also maximized if the second unit is also located in region two. Production is expanded in region two until marginal abatement costs fall to the level of marginal visibility damages; or, admitting other pollution effects at higher concentrations, until they fall to the level of marginal pollution damages in general. In this framework it is inevitable, from the standpoint of efficiency, that a form of zoning will arise as the indicated allocation. The production of income in one region, and its attendant increase in ambient concentrations and lowered visibility levels, raises the value of good visibility in the other region. Production facilities, embodying efficient abatement possibilities, will be concentrated in one region, while visibility values will be protected in the other. Thus, consideration of the substitution possibilities between regions reinforces the economic argument for a limited and dichotomous regulatory policy toward visibility protection.

Specialization between regions will not arise as the efficient pattern only if the more scenic region also happens to be the more productive, in the sense that more income can be generated per unit increase in ambient concentrations than in the other. Otherwise the marginal gains from interfering with visibility in the pristine region will always be less than those from expanding polluting production activities in the other.

Another look at column b of figure 13.4 shows that if it is worthwhile to protect visibility in other clean-air areas at all, it must be true that the slope of the marginal-abatement-cost curve is steeper than that of the marginal damage function. This provides guidance in the design of PSD policy to protect aesthetic values in areas other than those designated Class I areas that are afforded virtually absolute protection by the Clean Air Act visibility regulations (Weitzman 1974). If marginal damages changed more sharply than abatement costs, an emissions fee or penalty, to which individual emissions sources would adjust by abating pollution until their individual incremental costs equaled the fee, would be poor policy. A small error in estimating the penalty that would keep pollution within the safe limit would result in a large divergence of emissions from the optimum, because of the near constancy of incremental abatement costs. The sensible policy approach would be to fix an emissions or pollution limit at a safe level to prevent excessive damages, and then to allow the abatement costs to be determined by compliance with those limits. Because of the relative flatness of the cost curve, a small error in the emissions limit entails little penalty in extra abatement costs. Under these circumstances, a policy that implies an inelastic supply of entitlements is preferable.

At the other extreme, with an equal degree of uncertainty in costs and damages, if it is believed that marginal abatement costs increase rapidly over the relevant range, while the total damage function is about linear and marginal damages are nearly constant as pollution increases, then it makes little sense to adopt a policy that sets an inflexible limit on the amount of pollution, irrespective of costs. The wrong choice would involve either a large excess cost of abatement beyond the level justified by the damages the emissions would have caused, or large amounts of damage that could have been avoided by relatively inexpensive abatement. A policy approach that established limits on the costs of abatement, while letting the total level of emissions be determined by private decisions, would be more conducive to efficiency. An error in setting the penalty for emissions or the incentives for cleanup would not result in a large difference in the amount of abatement, because the abatement cost curve is elastic.

Of the two overall limits to degradation, area-wide emissions limitations allow tradeoffs among sources in adjusting emission levels whereas ambient limits determine inflexible limits to degradation at each location. Also, of the four policy options, a clear distinction can be made between those (the limited giveaway policy and the use of auction markets) that imply an inelastic supply of entitlements and those (the offset market with variable offset ratio and the offset market with price ceiling) that imply an elastic supply of entitlements. The last two admit the possibility that if cleanup costs get sufficiently high, enough entitlements might be supplied to infringe the overall limit. The last option, in fact, implies a ceiling on cleanup costs—a penalty at which the supply of entitlements becomes perfectly elastic. The offset market implies an elasticity that ranges from 0 to infinity as the ratio varies from 1 to 0. With these policies there is a safety valve on cleanup costs, but the level of emissions and environmental effects is left, in part, to the operation of market forces. With the first two policy options the maximum emissions that might ensue are fixed, but the costs of abatement are unconstrained and will depend on market conditions. All four alternatives imply that there is a minimum price for location in clean-air areas; namely, the cost of adopting BACT on new facilities. This is the minimal amount of cleanup any new sources can undertake (Roberts and Spence 1976).

The nature of benefits and costs in PSD areas, and the characteristics of the available policy options, imply that a more efficient policy under uncertainty would be one that provided entitlements to emit with considerable elasticity of supply rather than one that set rigid standards. The appropriate price at which entitlements were made available would be one reflecting as accurately as possible the marginal valuation of visibility in such areas. It is evident that, should it not be worthwhile to protect visibility, as in the situation depicted in graph i in column b of figure 13.4, no abatement would be undertaken. The charge p would be experienced as a dead-weight revenue loss by industry, and could, in principle, be rebated. If it should be worthwhile to protect visibility, then a program that sets limits on abatement costs is likely to be more efficient than one that defends certain standards at any cost. This conclusion, if we take into account the special protection given to scenic Class I areas by the visibility regulations of the Clean Air Act Amendments of 1977, reinforces the findings with respect to health risks and material damages. Because that damage function is undoubtedly quite flat in the range of concentrations found in PSD areas, policies that imply an elastic supply of entitlements seem more desirable. This leads to the following general conclusions:

• Area-wide emissions limitations seem to be a more appropriate means of limiting degradation than ambient increments or no limits, in view of the nature of the damages PSD policy is to prevent, the uncertainties of modeling, the behavior of plumes in rough terrain, and the probable differences in regional growth rates.

• With respect to the implications of the four allocation policies for air quality, incentives to abate, and patterns of economic growth and land use, it is difficult to arrive at generalizations for the entire United States. The most valid conclusion is that the relative suitability of the policy options will depend on the specific conditions in the regions to which the policies would apply.

• It is also difficult to generalize about the contribution each alternative approach would make to reducing the resource cost of defending overall emissions or ambient limits; that is, their contributions to cost-effectiveness. Aside from the conclusion that the present policy of limited giveaway always makes the least contribution, it is only possible to state that the relative contributions of the alternative approaches will depend on identifiable, and potentially measurable, characteristics of the region in which they might be applied.

• With respect to the likely contribution of the options to overall efficiency in regulation, it is possible to make a somewhat stronger statement. With the large uncertainties involved, it is likely that a twofold strategy might be desirable. One component is the strict visibility-protection policy to protect Class I areas, through siting restrictions if necessary. The other component would be a policy toward the supply of entitlements to emit in other PSD areas that implied considerable elasticity with respect to the abatement cost incurred. Of the four approaches considered in this analysis, the offset market with variable offset ratio, the offset market with price ceiling, and an emissions fee adapted for use in new-source permittings are of this type.

The conclusions offer considerable justification for PSD policies that differ from region to region in accordance with the nature of the air-quality values being protected and the likely characteristics of current emissions and future growth.

Chapter 14

Economic Incentives and Environmental Policy: Politics, Ideology, and Philosophy

Steven Kelman

To shed light on why economic-incentive approaches have had so little appeal in the political system, and to assess their prospects for greater political success, the most active Washington-based participants in environmental policy formation were interviewed in early 1978. Some of the results were startling. One initial hypothesis had been that a significant reason why economic-incentive proposals had not gotten further was that such proposals raised implementation problems that academic economists were not likely to have thought about. Many ideas that seem attractive end up being much less so because their implementation would require organizational capabilities that cannot be created. The assumption was that people closer to the day-to-day formulation of environmental policy would provide information on the "boring details" that so often stand between the grand visions of those conceiving new policy ideas and the real-world outcomes desired. However, the respondents did not supply any new reasons, garnered from their greater practical experience, to reject charge proposals as theoretically appealing but organizationally unachievable.

What the interviews did hint at, instead, was the extent to which proposals to use economic incentives for environmental policy raise ideological or philosophical issues about what kind of society we want to live in. The interviews showed that economic-incentive approaches did have supporters among participants in environmental policy making—especially among Republican congressional staffers. But these supporters did not, by and large, support charges because of their efficiency advantages. The interviews showed surprisingly little knowledge of the nature of—or even the existence of—the efficiency argument for charges. Advocates supported charges in significant measure because of a general ideological or philosophical conviction favoring markets and opposing "government interference." This conviction was based on the view that the market was a realm of freedom whereas government was a realm of coercion—a view connected with general opposition to many of the demands government has made on citizens to behave in ways they otherwise would not behave.

Opponents of charges, too, tended to be unfamiliar with the efficiency arguments. Those environmentalists (and their congressional allies) who were opposed to charges were partly worried for strategic reasons about the effect of moving from standards to charges on the political prospects for maintaining strong environmental laws. This included concern about the effect of moving jurisdiction over environmental issues from congressional environmental committees, which have many committed environmentalist members, to tax-writing committees, which were regarded as more conservative and less proenvironment. It also involved worry about the effect of reopening environmental laws to the revision that would be required to set up a charge system at a time when environmentalists were politically not as strong as they had been when the original legislation was passed in 1970 and 1972. However, for many environmentalists and their congressional allies, a key element in the reaction to charge proposals was (just as for supporters) a general ideological or philosophical attitude toward the market and toward government. Although many supporters of economic incentives endorsed the market, most environmentalists and their congressional allies were disturbed by it. The worries about charges being a "license to pollute," generally the object of scorn and condescension by economists, reflect a fear that an economic-incentive approach to environmental policy represents adoption of an attitude of moral indifference by society toward polluters. The attempt to channel behavior by changing incentives gained grudging respect for effectiveness, but aroused fear for its unconcern with the motivations for behavior. And there was uneasiness about the idea of placing yet another valued thing (in this case environmental quality) into the system of buying, selling, and pricing that the market represents.

These anxieties are partly related directly to the environmental concerns of those interviewed and partly related to broader social concerns they have. Part of the reason for the uneasiness about placing environmental quality into the market system is the fear that this would lower the perceived value of environmental quality for many people. This in turn would tend to lower the ambitions for environmental cleanup expressed in government policy—ambitions that are a function of the value people attribute to environmental quality. And, insofar as the anxieties express broader concerns, it would be out of place to suggest to environmentalists, or to anyone else, that it is inappropriate for them to have such broad concerns beyond the specific issue area they deal with. My own view is that these worries tap something real and important. This is why decisions to use economic incentives as a policy tool in any area of public policy, including environmental policy, cannot be evaluated on the basis of their efficiency advantages alone; such decisions may make the society less attractive in other respects.

The interviews revealed relatively little support among trade-association environmental lobbyists for economic-incentive proposals, except among some who were ideologically committed to markets. In fact, though charges may provide a cheaper way for society to meet given environmental goals, the private cost to industry of attaining these goals through charges might well turn out to be greater than the cost of attaining them through standards. This is because firms would have to pay the government a charge for each unit of pollution emitted above zero; the total cost to the firm then becomes the cost of reducing pollution levels plus the cost of the charge on residual units of pollution. With standards, nothing must be paid beyond the cost of reducing pollution to the level of the standard. A firm for which reducing pollution to the level of the standard was very expensive might save money by not having to meet the standard and abating pollution only to the point where the marginal cost of abatement equaled the charge, while paying the charge on residual units. However, for many firms the combined cost of abating to the level where the marginal cost of abatement equaled the charge and then paying the charge on residual units would be greater than simply abating to the level of the standard. (This cost of paying the charge for residual units of pollution is a private cost to polluting firms. However, to an economist charge payments are not a social cost, because such taxes do not use up real resources but only transfer them to the government, which can then distribute them in various possible ways.)

The interviews revealed a lower level of knowledge about economic-incentive proposals among industry lobbyists than among any other group surveyed. It would therefore be incorrect to conclude that industry opposition was based on a rational calculation that the private costs of charges could well be greater than the private costs of standards. Yet the relative ignorance of most industry lobbyists interviewed relates to industry's tendency, as the party opposed to greater governmental incursion into their affairs, to react to proposals others make for new government action rather than initiating proposals of its own. (This has certainly changed somewhat since the interviews were conducted, and especially since the accession of the Reagan administration.) One may assume that if proposals for charges became politically serious, trade associations would become seriously knowledgeable about them and would discover them to be to the financial disadvantage of most firms. At that point, industry opposition, except among a minority whose ideological commitment to markets outweighed their self-interest, would likely become both well-informed and vocal. Proponents of economic incentives have, then, little prospect with industry; the more seriously their proposals are taken, the more vocal industry's opposition is likely to become.

The hope that proponents of charges have for expanding political support beyond Republicans who are ideologically committed to markets must, then, lie with environmentalists and their Democratic allies. These groups cannot oppose, *ceteris parabis*, getting more "bang" for the environmental buck. But it is among these groups that the deeper anxieties the use of economic incentives raises are felt.

If an environmentalist is worried about the strategic implications for the maintenance of strong environmental laws of a move from standards to charges, and if he is uneasy about the market for some general ideological or philosophical reasons, he is certain to oppose economic-incentive proposals unless he perceives a strong argument that there are efficiency advantages in being able to obtain a given degree of environmental quality more cheaply using charges than using standards. What would then face environmentalists would be a tradeoff among valued but conflicting goals. The judgment would have to be made whether the efficiency advantages of charges outweigh their strategic, ideological, and philosophical disadvantages.

The Economists' Case for Incentives

The two most widely read arguments in favor of the use of economic incentives for environmental policy in particular and for public policies to deal with the external effects of individual behavior in general are *Pollution, Prices, and Public Policy* by Allen Kneese and Charles Schultze (1975) and Schultze's *The Public Use of Private Interest* (1977). Looking at them allows one to analyze the messages being sent to environmental policy makers by supporters of incentive approaches to see what kinds of arguments they were making. Kneese, an environmental economist, has testified at congressional hearings, spoken repeatedly with the staff of Senator Peter Domenici (who represents Kneese's home state of New Mexico and sits on the Senate's environmental affairs subcommittee), and made himself well known among environmental policy makers in Washington. Schultze was director of the Bureau of the Budget under Lyndon Johnson and became, over the years, well known among Democratic economists. He served as chairman of the President's Council of Economic Advisors under Jimmy Carter.

It is striking, in examining these two books, to note how much they differ in emphasis. In the case for economic incentives that Kneese and Schultze make in *Pollution, Prices, and Public Policy*, efficiency arguments hold pride of place. The authors do argue that charges will be easier to enforce than standards and that they promote technological innovation, and there is also a brief section entitled The Broader Problem of Incentives where the authors note that in government policy "little effort is put into devising new incentives or correcting existing ones to spur individual decision makers, in their own self-interest, toward socially desirable actions" (p. 7), but most of the book deals, not with broader issues involved in the use of economic incentives in environmental policy, but with the unnecessary costs that a standards approach imposes. In introducing "the basic economics of pollution control," Kneese and Schultze note right off (p. 19) one of the "basic facts" about the economics of pollution: "The costs of pollution control vary substantially from industry to industry and from firm to firm. To be efficient, the degree by which individual sources have to reduce their pollution discharges should vary in relationship to the cost of production." In the course of the book Kneese and Schultze present a fairly detailed exposition, in language accessible to noneconomists, of why charges allow attainment of a given degree of environmental quality at a lower cost than standards (pp. 85–91). They present estimates of potential cost savings achievable through allowing differential reduction across sources, and they conclude that "the difference between inefficient and efficient control policies can mean scores, perhaps hundreds, of billions of dollars released for other useful purposes over the next several decades" (p. 83).

In *Pollution, Prices, and Public Policy* Kneese and Schultze link their case for charges with a plea for greater consideration of the enormous cost of attaining very high total cleanup levels. The first of the "basic facts" about the economics of pollution that they introduce is that "the cost of reducing any type of pollution increases more than proportionally with the amount of pollution removed" (p. 19). The gist of the subsequent discussion—that "going from, say, 97 percent to 99 percent removal may cost as much as the entire effort of going from zero to 97 percent"—suggests that decisions to aim for very high total cleanup levels are unwise. Later on, Kneese and Schultze garble messages regarding efficiency and marginal cost when they write in one sentence about "the importance of designing efficient programs" and "the need to evaluate environmental goals in terms of costs" (p. 71).

There is no logical connection between the efficiency argument for charges and the argument that very high cleanup levels impose marginal cleanup costs that may not outweigh marginal cleanup benefits. Charges would produce the most efficient way of attaining any degree of cleanup chosen, and the decision about charges versus standards is separate from the decision about what cleanup level to aim at. (It is true that meeting a very high overall cleanup level might necessitate a charge so high that virtually all polluters would find it advantageous to reduce emissions dramatically. This in turn would reduce, or at the extreme of zero discharge even eliminate, the efficiency advantages of charges.) But though there is no logical connection between the two arguments, they are analytically linked by both emerging broadly out of the microeconomic paradigm. And they are politically linked in that economists, who have sought to get proposals to use economic incentives onto the political agenda, have also been among those who have worried that the costs of the level of cleanup environmentalists have demanded are not worth the benefits. This cannot be expected to help the cause of economic incentives with environmentalists.

The brunt of the case in Schultze's 1976 Godkin Lectures at Harvard, published shortly thereafter in book form, was different. The title, *The Public Use of Private Interest*, summarizes the thrust of the argument. "The basic theme of this book," Schultze writes, "is that there is a growing need for collective influence over individual and business behavior that was once the domain of purely private decisions." This is because, in a complex society, the external effects of our behavior increase as our interactions with one another increase. But, Schultze continues, "as a society we are going about the job in a systematically bad way" because "instead of creating incentives so that public goals become private interests, private interests are left unchanged and obedience to the public goals is commanded" (pp. 5–6).

The Public Use of Private Interest removes efficiency from the head of the parade, where it stood in *Pollution, Prices, and Public Policy*. The word *efficiency* is scattered throughout, but Schultze never presents a coherent argument, for someone who did not already understand what he was talking about, explaining why economic-incentive methods are capable of achieving a given goal at less cost than the "command and control" regulation he criticizes. Indeed, Schultze's discussion of the advantages of the market as a form of social organization excludes efficiency entirely. It is mentioned only in a dependent clause in a sentence beginning "Quite apart from the maximizing characteristics elaborated in formal economic theory, . . . " (p. 16). Instead, Schultze's case for the market is predominantly a philosophical or ideological one. Its first pillar comes squarely out of a philosophical tradition that grew from Adam Smith's notion that individual pursuit of self-interest would, in a regime of competitive markets, maximize the social good. That tradition is so firmly embedded in economics by now that most economists probably do not realize, unless they venture out into the world of noneconomists, that it is a proposition of moral philosophy far more controversial than the self-evident proposition that it is preferable to achieve a given goal at a lower than at a higher cost. "Harnessing the 'base' motive of material self-interest to promote the common good," Schultze writes, "is perhaps *the* most important social invention mankind has yet achieved" (p. 18). It might be nice if human activity could be organized on other principles, but "however vital they may be to a civilized society, compassion, brotherly love, and patriotism are in too short supply to serve as substitutes." The advantage of arrangements that can turn self-interest into social advantage is that they make use of motivation whose supply is not subject to doubt. They are thus more likely to achieve the desired results than are arrangements relying on other motivations: "If I want industry to cut down on pollution, indignant tirades about social responsibility can't hold a candle to schemes that reduce the profits of firms who pollute. If I want drivers to economize on gasoline usage, advertising appeals to patriotism, warnings about the energy crisis, and 'don't be fuelish' slogans are no match for higher prices at the gas pumps." What is needed is not moral condemnation, but changes in incentive systems: "In most cases the prerequisite for social gains is the identification, not of villains and heroes but of the defects in the incentive system that drive ordinary decent citizens into doing things contrary to the common good."

Although this argument's philosophical roots in the Smithian tradition are important, it should be noted that the case Schultze is making differs somewhat from Smith's. Smith argued that pursuit of self-interest in a regime of competitive markets would automatically, without further social intervention, maximize the social good. But that argument breaks down when an individual need not consider the external costs of his actions, in which case an individual's self-interested acts can damage others. If these damages are greater than the gains the acts bring the individual, the net social good is not maximized. Schultze's book deals with such cases of external effects, and he realizes that the process of relating pursuit of self-interest to maximization of the net social good no longer has the automatic quality described by Smith. Social intervention, Schultze realizes, is necessary to make public use of the private interest. But the general judgment in favor of accepting pursuit of self-interest in organizing social arrangements, instead of trying to combat it as was the traditional approach in religion and moral philosophy, remains to inform the choice Schultze makes about how best to deal with external effects.

The second pillar of Schultze's philosophical case is that markets offer people greater choice than government commands. "While some element of coercion is implicit in any social intervention," Schultze writes, "the use of market-like incentives to achieve public purposes minimizes that element" (p. 6). The first advantage of a market arrangement Schultze enumerates is that it is "a form of unanimous-consent arrangement." Government decisions, by contrast, "necessarily [imply] some majority who disapprove of each particular decision." Market arrangements thus "minimize the need for coercion as a means of organizing society" (pp. 16–17). The most sustained efficiency-related discussion in Schultze's book involves the difficulty government has gathering enough information about complex problems involving huge numbers of citizens to be able intelligently to formulate regulations to change behavior. Yet even this argument is couched in terms of greater freedom to choose. We worry about the relationship between educational inputs and learning outputs, and gather lots of information to try to answer such questions, Schultze argues, because education is not sold on the market. We need not do the same, he continues, with recreation: "Because individuals can choose freely, the prices that millions of them are willing to pay reflect the values they put on various forms of recreation." Free choice based on individual valuations determines the level and mix of recreation production, rather than a political decision that requires a costly information search (p. 19).

The Interviews

The criteria for selecting respondents for the survey were simple. We selected

every staff member (Democratic and Republican) of the Subcommittee on the Environment of the Senate Environment and Public Works Committee, the Subcommittee on Water Resources of the House Public Works and Transportation Committee, and the Subcommittee on Health of the House Interstate and Foreign Commerce Committee,

personal-staff members of the relevant committees who reported spending a "significant" amount of their time on environmental policy (not all the House Subcommittee members had personal staffers devoting any significant time to environmental policy), and

all the professional staffers of the Washington offices of environmental organizations and trade associations, identified in a publication provided by the EPA public affairs office, who dealt with air and water pollution (and, for environmental groups, solid waste as well).

No executive-branch officials, in either the White House or the EPA, were interviewed. We sent letters explaining the project and called to arrange interviews. In order not to bias replies, respondents were not informed in advance of our specific interest in attitudes toward incentive approaches. Nobody refused to see us. The total number of respondents was 63. This number is too small to make possible any sophisticated analysis of the data, especially when it is (as it will generally be in the presentation to follow) divided up by respondent group. The data will therefore be displayed as simple marginals or, occasionally, in twofold cross-tabulations.

The interviews generally lasted between 45 minutes and two hours. Almost every question was of a fixed-format, open-ended type. Because of the length of the interview and the differing amounts of time that different respondents had free, in most cases not every question was asked every respondent. There was no tendency, however, for the more "important" respondents to give us less time than other respondents; if anything, the opposite may have been the case. The marginals and cross-tabulations to be presented, therefore, often are based on a subset of those interviewed.

Questions in the first third of the interview dealt with the respondent's background and solicited information about the respondent's major criticisms of existing environmental policy, ideas for new approaches or directions for environmental policy, and sources of information for

Table 14.1   Attitudes toward charges (percentages of entire sample; $N = 61$).

For	23
Against	57
For experiments (in areas currently unregulated, in one region) but against full implementation	17
For, but only with very gradual implementation	2
For a hybrid system of charges and standards	2

Note: Percentages add up to more than 100 because of rounding.

new policy ideas. These questions were asked before any of the questions that focused on economic incentives. The idea was to see what concerns the respondent expressed spontaneously before the subject of the interview turned to charges.

The survey showed that nearly every respondent had heard of proposals to use charges in environmental policy; only two of the 63 respondents (both on the personal staffs of congressional environmental committee members) had not. Furthermore, virtually all respondents, in response to the question "Could you explain why some people regard charges as superior to standards?," were able to cite some arguments that proponents of charges make in favor of such proposals. And the responses showed that support for a charge approach was not insignificant. For the sample as a whole, table 14.1 displays responses to the question about whether they favored introducing charges to replace regulation using standards.

Two things immediately stand out in the survey responses taken as a whole. The first is the pattern of support for and opposition to economic-incentive approaches. The most sympathetic group was Republican committee staff. Of the four Republican staffers of the Subcommittee on the Environment of the Senate Public Works Committee, three supported moving to a charge approach. All three, in fact, spontaneously mentioned failure to use economic incentives in the questions on criticisms of previous environmental policy and new approaches to environmental policy. The most united opponents of charges were the Democratic committee staffers. The five Democratic counterparts to the Senate Republican committee staffers were all opposed to economic-incentive approaches. (The one Republican Senate staff member interviewed who opposed charges said during the interview that he had been a Democrat when hired to work for a Republican senator on the committee; he had decided to go to Washington on the recommendation of a professor in the environmental-law program where he was studying, and the job with this Republican senator from his home state was the best one available when he looked.) The 19 environmentalists interviewed, by contrast, were split; 32 percent were classified as favoring charges, 16 percent as favoring experimenting with them on a limited basis to see how they worked, and 37 percent as hostile. The 20 industry environmental lobbyists were mostly opposed to charges; 15 percent were classified as favoring them and 85 percent as opposed.

This pattern of responses—with the exception perhaps of the industry hostility—appears to suggest a liberal-conservative ideological tenor to the reactions of respondents. This impression is dramatically confirmed by an examination of answers to the question where respondents were asked to cite arguments that economic incentives proponents made in favor of charges. What stands out dramatically is how few respondents in any of the groups surveyed cited the efficiency argument. In the sample taken as a whole, only 16 percent spontaneously cited the efficiency argument. (To be counted as having cited the argument, the respondent need not have used the word "efficiency," but need only have explained even somewhat the idea of cost minimization in achieving a certain goal.)

Respondents who did not spontaneously mention the efficiency argument for charges were specifically asked two followup questions. The first was: "The argument is sometimes made that using charges we could obtain any given degree of environmental quality for a lower cost than by using standards. Have you ever heard this argument for charges?" Sixty-four percent of the respondents who had not spontaneously mentioned the efficiency argument earlier ($N=36$) claimed to have heard the argument. Yet, when then asked "Could you explain on what basis people make this argument?" not a single one was able to do so. Fifty-five percent ($N=43$) gave no explanation at all, and 44 percent gave arguments that related to a general ideological case for charges ("Costs less because less bureaucracy costs less"; "Costs less because the regulated parties have more information and experience than the regulators"). A few of those who were unable to explain the efficiency argument made reference to having seen arguments economists made in support of the contention, but added frankly that they were unable to understand the argument ("I've seen graphs and crossing diagrams"; "Lots of economists talk a lot of gobbledygook that the average layman can't understand"). Respondents were also asked, in the section of the survey preceding the explicit question on charges, "If you think of different sources of a given type of pollutant, is it your impression that different sources tend to face relatively similar costs in attaining any given proportion of pollution abatement or very different costs?" A large majority of respondents were aware that different sources often faced very different costs, a key factual component of the efficiency argument for charges. (The percentage was 89 ($N=45$) for all respondents, with no difference between industry and nonindustry respondents.) Yet only two of the 24 respondents asked mentioned charges in answering the followup question, "How would you go about dealing with this problem of differing costs?" Thus, although the survey showed that economic-incentive proposals most definitely had advocates, the efficiency argument in support of such proposals appeared to have passed most respondents by.

Instead of technical arguments about efficiency, arguments best described as ideological had made the greatest impression on respondents (with the exception of those in industry, to whom I will return shortly). By "ideological" arguments, I mean arguments reflecting a general attitude that "the market" (or "the government") is a good or bad way to go about solving problems, buttressed by reasons not related to the efficiency arguments for markets. Proponents of charges were endorsing, in a general ideological way, "the market," and excoriating government and bureaucrats; opponents of charges were uneasy about or hostile to "the market" and more convinced of the necessity for the government, bureaucrats and all. This tendency appeared most clearly in the responses of the congressional environmental staffers, particularly the Senate Democratic and Republican staffers. Table 14.2 displays the numbers of respondents who mentioned different arguments for charges in response to a question about what arguments supporters of charges made in support of such proposals. (There were only five Democratic Senate staffers on the subcommittee and four Republican staffers. Percentages should therefore be interpreted accordingly. Also, percentages add up to more than 100 percent because a respondent could mention more than one argument.) Table 14.3 displays responses to the followup question, "In your own view, what are some of the disadvantages, if any, of charges over standards?"

Democratic Senate subcommittee staffers were hostile to charges, whereas Republican Senate subcommittee staffers were enthusiastic. But when asked to cite arguments generally made in favor of charges, both groups emphasized ideological arguments ("less bureaucracy," "uses the market," "appeals to self-interest"). When asked to cite what they saw as the disadvantages of charges, the hostile Democratic Senate

Table 14.2 Arguments for charges cited by Senate committee staffers.

	Percentage of respondents who cited argument	
	Democratic staff ($N=5$)	Republican staff ($N=4$)
Charges make it in one's self-interest to reduce pollution	60	25
Case of enforcement	60	0
Less bureaucracy; simplicity	40	100
Promarket or antigovernment comment without further justification; general "prochoice" comment	20	75
Correctly stated efficiency argument	20	25
Stated efficiency argument but did not understand it well	20	0
Incentive to go beyond a standard	20	25
Internalized costs; "polluters should pay"	20	25
"Based on cost-benefit considerations"	20	0
Charges encourage technical innovation	0	25

subcommittee staffers also tended to give ideological objections: "shouldn't have a choice to pay and pollute," "license to pollute," "can't put a monetary price on health or the environment." Some Democratic staffers were also worried about the political difficulties of setting charges high enough, particularly at a time of less interest in the environment than during the passage of environmental legislation. By contrast, the main problem the Republican staffers mentioned was the technical difficulty of setting charges—a narrowly focused, non-ideological concern that not a single Senate Democratic subcommittee staffer mentioned.

The most important figure in the formulation of environmental policy in the United States in the 1970s was Senator Edmund S. Muskie, then chairman of the Subcommittee on the Environment. The major occasion when Muskie spoke publicly on charge proposals was in a Senate floor debate in November 1971 with Senator William Proxmire, who had introduced an amendment to pending clean-water legislation calling for the use of effluent charges. (Proxmire had been influenced by environmental economists on his staff.) "We cannot give anyone the option of polluting for a fee," Muskie stated (*Cong. Rec.*, pp. 38,826–38,834). In the floor debate, Proxmire never presented any efficiency arguments for charges. Instead, he argued that charges would be more easily en-

Table 14.3 Disadvantages of charges cited by Senate committee staffers.

	Percentage of respondents who made argument	
	Democratic staff	Republican staff
One should not have a choice of polluting if willing to pay a charge; "license to pollute"	80	25
Health damage should never be allowed; one cannot set a monetary value on health or the environment	80	0
Political environment now is less in favor of environmental controls; reopening the issue risks ending up with looser standards	40	0
Politically difficult to set charges high enough	20	0
Administrative problems; difficult to monitor; too much bureaucracy	20	25
High cost of changing from the present system, given that system is in place	20	25
Industry could just pass on costs and not control pollution	20	0
Charges difficult to set; technical and economic information needed	0	100
Results are uncertain	0	25
Charges inequitable (would not allow variances); would hit the poor companies too hard; would concentrate industry	0	25

forced, with less bureaucracy—a notion Muskie rejected. The debate included frequent references from both sides to "the cost of cleaning up," as if that cost were constant across different pollution sources. And when the subject of charges again came up in a Senate hearing in 1977, when Senator Gary Hart discussed economic incentives favorably (also emphasizing that they would be less bureaucratic), Muskie again rejected the notion that charges would involve less bureaucracy.

The two most important Democratic staff people in Congress dealing with the environment at the time of the survey were Karl Braithwaite and Leon Billings. Braithwaite had only recently become director of the subcommittee's professional staff; he had taught political science and environmental policy at Duke University before coming to the committee in the early 1970s. Billings had worked for Muskie since the mid-1960s and was majority staff director when air and water legislation was written in the early 1970s.

In a brief interview, Billings began by stating: "There is a basic philosophical difference between regulatory people and economists. Economists don't care whether you achieve a reduction of pollution. They don't really care, but we really do care. The economic approach is inequitable, because it permits a choice on the part of the polluter whether to spend on pollution control or on other things." Billings went on to say that the problem with economists was that "they worship things economic." Billings stated that he had "heard" the efficiency argument and added that "it may or may not be valid." He was not able to explain it. Instead, he used the question to express distaste for economists, whom he regarded as "zealots."

Braithwaite gave a lengthy interview that mixed strategic with philosophical opposition to charges. "The biggest argument the proponent of charges made is that you get an automatic correctiveness by drawing in more market forces. They love the market, and they want to readjust the equation slightly for externalities so that market can go ahead and run." He said that another main argument for charges was that they embodied a cost-benefit approach, "since you're never justified in setting the charge at more than what cost-benefit analysis tells you is justified." He continued: "All we ever hear about from industry is about costs. Our members, though, listen to people talking about values they want to protect—nature and parks and health. We had a guy who testified here who was ultrasensitive to pollution. The committee has to decide whether we want to protect a guy like this. How do you quantify that?" Braithwaite went on to say that using charges to deal with solid waste was more acceptable than for air and water "because there is nothing that hurts your health about an excessive candy bar wrapper." Finally, Braithwaite also rebutted the argument, which he thought was an important part of the case for charges, that they would mean less bureaucracy.

Further consideration of the larger belief structure of which these views of Senate Democratic staffers form a part which will be presented below in the section on environmentalists. Quotes from the interviews with the other three Democratic staffers, however, give a flavor of their concerns:

The best argument they make is that industry works on the basis of the profit motive and that they will respond best to a pollution-control system which uses the same motive.

The philosophical assumption that proponents of charges make is that there is a free-market system that responds to pressures and that responds to relative costs—a marketplace that is responsive. I reject that assumption. I grew up with Lockheed loans and auto emission price-fixing and cost pass-ons.

The argument for pollution charges is that since we do have a capitalist framework, an economic driving force is more effective. . . . But I'm troubled by the psychology of it. If there's a national program, that says we want every corporation to become part of achieving that goal. To allow them to buy their way out is not good national psychology. It's like buying your way out of the army. If you're rich enough to pay a gas-guzzler tax, you can buy your way out of our national policy.

The things I want to attain with environmental policy aren't quantifiable. They can't be measured in dollars. These values can't be preserved if the calculation is essentially a business calculation. . . . The people I have spoken with about charges were mostly academicians. They struck me as being ignorant of environmental values. I can't imagine writing off 300,000 people in Los Angeles. I don't know how much I'd be willing to pay, but it's lots more than we're talking about. The Clean Air Act was written to protect vulnerable segments of the population. . . . I'd be more willing to see a charge approach for solid waste. There's an economic value there in helping cities save money in managing solid wastes, and that's more compatible with an economic approach than in the case of air and water.

What was perhaps most striking about the arguments of the Republican staffers was the failure of two of the three supporters of economic incentives to present the efficiency argument for charges. The senior Republican staff member for the subcommittee stated spontaneously early in the interview that "philosophically the market mechanism has greater appeal" than regulatory approaches. When asked about arguments proponents made in favor of charges, he did correctly state the efficiency argument, but only after stating that the first argument for charges was that it would "get decisions made by the guy who's going to pay rather than by the bureaucracy." After discussing problems he saw with getting charge proposals through Congress and setting charge levels, he concluded that on balance he was for using charges "because I trust the marketplace more than the bureaucracy."

Two other Republican subcommittee staff members were more enthusiastic—and also more ideological. Both spontaneously mentioned economic incentives early in the interview. Neither saw any significant drawbacks to charges, but both thought that charges would be cheaper than standards only because the bureaucracy would be taken out of things. "Charges will be cheaper for a similar level of control, because there would be less administrative costs." said one. "You could get a whole lot better working out in the market, without bureaucratic enforcement," said the other. "Things take place independent of a big structure." As for the arguments for a charge system, it was first mentioned that it "relies on market mechanisms rather than bureaucrats having to make judgments," that "market forces make the determination instead." Both respondents gave similar responses when asked who the main opponents of charge proposals would be. "People who like government meddling," was one response. "People disenchanted with the marketplace and private enterprise" was the other. The current regulation strategy had come about because of "disenchantment with the marketplace and private enterprise" and "a desire to kick industry in the ass."

Of the other congressional staffers, those most similar to their Senate subcommittee counterparts were Democratic staffers for the Subcommittee on Health of the House Interstate and Foreign Commerce Committee. (This subcommittee had no Republican staffers.) The Interstate and Foreign Commerce Committee is the site of a great many ideological battles (over health-care issues, notably, as well as air pollution). One leading staffer on the House subcommittee expressed an ambivalence toward charges that, he stated, related to his ambivalence about what attitude to take toward "bad" behavior such as that of polluters. He was afraid that charges "might relegate social policy decisions to the polluters." Health, he continued, "is a moral issue, and I'm afraid of putting the fox in charge of the chicken coop by leaving this to the judgment of the private sector." Asked later in the interview whether, if a charge were put on driving a car downtown that reflected the social costs of such driving, he would still criticize somebody who drove his car and paid the charge rather than taking available public transportation, he replied: "I have mixed feelings. It has to do with my own personal political migration. It used to be that I would have criticized such a person harshly. Now I have some sympathy for allowing him the freedom to do what he wants. But my migration isn't complete. There's still something within me that wants people to do the right thing."

By contrast, the House Committee on Public Works and Transportation is traditionally a fairly conservative, Southern-dominated committee that oversees many "pork barrel" projects in addition to its jurisdiction over water-pollution legislation. The staffers on that committee, Democratic and Republican, appeared to be somewhat less ideological and to be concerned about charges as much for their potential to disrupt the program as for any other reason.

The personal staffs of senators who were members of the Subcommittee on the Environment, and even more so the personal staffs of House members (who must subdivide their time among many more issues), were generally much less knowledgeable than the committee staffs about issues. (It was from among this group of ten that the only two respondents who had never heard of charge proposals were found.)

Gary Hart, a first-term senator from Colorado, was the only Democratic member of the Subcommittee on the Environment, at the time of the interviews, sympathetic to economic incentives in environmental policy. (During the 1977 hearings on revisions of the Clean Water Act, Hart frequently asked witnesses questions about their attitudes to charge proposals. He also proposed in 1977 using a tax on nitrogen oxide auto emissions as a way of breaking the deadlock between the auto industry and environmentalists on whether to relax auto emissions standards written into the Clean Air Act amendments of 1970.) Important in this context was the apparent motivation for Hart's stand. A product of the antiwar protest movement of the 1960s, Hart had managed George McGovern's presidential campaign in 1972. The "new politics" movement out of which Hart emerged differed from traditional American liberalism in its relative skepticism about the benevolence of government, nurtured by opposition to the war in Vietnam, and a greater concern for "quality of life" as opposed to New Deal welfare-state issues. In the political climate of the late 1970s, such concern became translated into attempts to develop a new political approach that mixed "liberal" concern with issues like the environment or tolerance of unconventional life styles with "conservative" hostility to government intervention in people's lives. Hart, who was once quoted as stating that the new generation of liberals didn't want to be "little Hubert Humphreys," became one of the spearheads of this new amalgam among younger members of Congress.

Advocacy of economic-incentive proposals for environmental policy fitted exquisitely into this vision, and Hart embraced such advocacy warmly as soon as the notion was presented to him. But, again, the reasons had much more to do with general ideological concerns than with efficiency. Hart presented his views at a Senate hearing with EPA Administrator Russell Train:

I see the country approaching . . . what I consider to be a small crisis in our government's entire regulatory approach to human conduct. . . . I . . . feel we have almost run out of steam on the regulation of human conduct. . . . [We] must begin to think very seriously about substituting for this regulatory approach the economic-incentive approach. It is not foolproof. It is not perfect. But all I am saying is that I think the time is rapidly approaching when the citizens of this country are just going to routinely toss out of office people who continue to support more regulation in their life. . . . It is stifling people's freedom. (*Status of the Programs*, pp. 39–40)

One of Hart's two environmental aides was able to state correctly the efficiency argument for charges; the other was not. Both, however, emphasized other things. The "politically sexy side" of charge proposals, one of the aides stated in an interview, was that "political demagogues are getting elected on the basis of less government interference with your lives." Charges were a way to have your environmental cake and still be able to answer those concerns about governmental interference. Charges would encourage polluters to lower pollution levels even below those set by standards. Using charges, the aide continued, "we may be able to get a better level of environmental quality that would be politically infeasible with standards." "Complying with the standards isn't enough," he said. "We want to get lower than that." And charges would allow people to decide for themselves. "If you can get better results than the existing approach and get the government off your back, you have a winner," the aide concluded. In response to a later question on parking bans versus transportation surcharges, the same aide responded that he favored surcharges over bans: "With a fee, a person makes a decision himself. People ought to make decisions themselves. They're happy when they do. . . . A charge allows more options. Parking bans allow fewer options."

As a group, environmentalists were by far the most knowledgeable about charges—and also the most split. Table 14.4 shows the arguments cited by environmentalists for charges. Thirty-two percent cited the efficiency arguments—not an overwhelming percentage, but impressive compared to the 0 percent of industrial respondents and the 10 percent of the congressional staffers. Table 14.5 shows the disadvantages environmentalists saw.

With the exceptions of the frequently mentioned efficiency argument and the "internalize costs" argument, the arguments environmentalists cited in favor of charges were quite similar to those cited by Senate staffers. So were the disadvantages they saw. In other words, there was a significant element of general philosophy or ideology.

The deeper structure of thought into which the worries of environmentalists and congressional majority staffers fit has several elements.

Table 14.4   Arguments for charges made by environmentalists ($N=19$).

	Percentage of respondents who made arguments
Correct statement of efficiency argument	32
Charges make it in one's self-interest to reduce pollution	32
Internalize costs; "polluters should pay"	32
Less bureaucracy, more simplicity	25
Case of enforcement	16
Knew no arguments	16
Promarket or antigovernment comment without further justification (prochoice)	11
Incentive to go beyond a standard	5
Uniformity	5
"Based on cost-benefit considerations"	5
Arguments for noncompliance fee only	5

Table 14.5   Disadvantages of charges cited by environmentalists ($N=19$).

	Percentage of respondents who made argument
One should not have a choice of polluting if willing to pay a charge; "license to pollute"	37
Health damage should never be allowed; one cannot set a monetary value on health or the environment	37
Politically difficult to set the charges high enough	26
Charges difficult to set; technical information needed; economic information needed	16
Results are uncertain	16
Difficulty in going from ambient health response to emissions charge	16
High cost of changing from the present system	11
Industry could just pass on costs and not control pollution	11
Charges inequitable (would not allow variances); would hit the poor companies too hard; would concentrate industry	11
Political environment now is less in favor of environmental controls; reopening the issue risks ending up with looser standards	5
Industry uncertainty	5

Table 14.6 Views on charges among environmentalists and Senate majority staffers mentioning various disadvantages.

	Percentage for charges	Percentage against charges	Percentage against full implementation; for use of charges in unregulated areas or experiments only
Mentioned "can't set price" argument (N = 1)	14	71	14
Mentioned "license to pollute" argument (N=7)	17	33	50
Mentioned neither argument (N=8)	50	38	12

These people were not trained philosophers, and their thoughts were not fully worked out. (An extended philosophical discussion of these issues appears in Kelman 1981.)

The most frequently expressed concern about economic incentives was that they would grant an unacceptable "license to pollute" and that one cannot set a monetary price on health or the environment. Thirty-seven percent of environmentalist respondents mentioned the "license to pollute" argument, and those mentioning that argument were also more likely to oppose charges (table 14.6). Economists discussing objections to charges have had trouble suppressing contempt for the "license to pollute" contention, which they interpret merely as an argument that, faced with a pollution charge, polluters simply will pay the charge and continue to pollute as much as before. Such an argument, in their view, demonstrates ignorance of the fundamental economic insight that if something becomes more expensive to do, less of it will be done. To the extent that they are willing to accept the "license to pollute" vocabulary at all, economists would give the counterargument that license fees are low charges, too low to diminish by much the behavior in question. If the pollution charge is set at an appropriately high level, it is continued, it certainly will decrease polluting behavior.

Some of the Democratic staffers and environmentalists who raised the "license to pollute" argument appeared to be mainly worried about whether the charge would be set high enough to decrease polluting behavior significantly, and it may well be that some staffers or environmentalists were ignorant of the insight that raising the price of a behavior means that less of it is done. But most of the unease about charges being a "license to pollute" reflected concerns that are not addressed by simply arguing that the charge can be made higher. Economics is explicit about its inability to contribute anything to discussions of what preferences people ought to have. Furthermore, mathematical proofs of propositions about the ability of competitive markets to maximize social welfare also depend on the assumption that the individual is the best judge of his own welfare and that individual preferences be taken as givens. These two factors, combined with the philosophical origins of contemporary neoclassical microeconomics in Benthamite utilitarianism, have produced in most economists a philosophical attitude of radical nonjudgmentalism about people's preferences. To most economists, as to Bentham, "the amount of pleasure being equal, pushpin is as good as poetry." (This attitude flows logically neither from the inability of economics to say anything about preferences nor from the requirements for mathematically deduced conclusions about conditions for the maximization of social welfare, but adopting a radically nonjudgmental stance does allow economists to claim more for the real-world value of their professional tools than would be the case otherwise.)

Licenses are normally given out to authorize behavior toward which society takes a positive, or at least neutral, attitude. Practicing medicine, driving a car, and getting married are not regarded as undesirable activities. One needs a license to perform these behaviors in order to assume that those performing them are competent enough to meet certain legal requirements. In other cases, as with licenses to own a dog, the behavior is also approved and the main purpose of the license requirement is to raise revenue. What many of those who worry about charges being a "license to pollute" react against is the implied authorization of polluting behavior that charges represent. They fear that by replacing standards (which are social statements of acceptable and unacceptable behavior) with charges, society would be saying, in effect, that "it's OK to pollute as long as you pay a fee." Hence such comments as the following:

When you have a manipulative approach like this, it isn't a societally stated goal to clean up rivers.

I think there's something anomalous about taxing a poison. A tax appears to sanction the poisoning.

How many people would feel that by paying their money they're absolved of guilt?

People should not pollute. The decision should not be up to them.

I don't like the idea that pollution is a matter of choice. You shouldn't be able to pay for it.

A guy living in the suburbs, driving a big car, using luxury appliances, should be blamed. It is morally wrong. There are alternatives. People can use carpools. People don't need electric can openers. There's a clear moral issue there.

I don't think anyone has the right to go out and ruin our environment.

Nader uses the words "environmental violence." I think he has a point.

One day a courageous district attorney will prosecute these people for murder.

The demand for charges reflects the deep resentment people had at being called criminals. They want to put this on a businesslike basis. But I'm not ready to abandon that approach.

Few environmentalists are willing to be nonjudgmental. They condemn polluting behavior. Indeed, it is at the heart of being an environmentalist to believe, and to propagandize for the belief, that one ought to have preferences that give a clean environment a strong weight. Statements by environmentalists frequently emphasize the importance of developing an "environmental ethic," a set of preferences that weighs environmental protection highly. In response to an early question in the survey on what criticisms they would make of recent environmental policy, 22 percent of environmentalists and Senate Democratic staffers ($N=23$) spontaneously mentioned a response coded as "low popular environmental consciousness." This frequency of response takes on added meaning as an indication of the topic's importance because the question talked about "environmental policy" and thus would appear to summon forth responses about government policy rather than individual preferences—hence the hesitation against a system for determining environmental quality that fails to make a social statement of disapproval about polluting behavior.

An economist might answer by agreeing that individual behavior that does not take into account its effects on others—such as polluting behavior absent a charge that reflects its social costs—is inappropriate. Indeed, that is why economists favor placing a charge on pollution in the first place. But if a charge is set at a level that does reflect the full social cost of the behavior in question, economists' nonjudgmentalism returns in full force. If the charge were to be paid, few economists would express any criticism of a person undertaking the behavior. But the attitude of many Democratic staffers and environmentalists went beyond that. Two questions were asked to explore such attitudes:

• Let's say that a parking surcharge developed as part of a transportation control program reflected the costs a driver imposes on society by driving a car, including the damages from auto pollution. If the surcharge reflected all such costs, would you then feel it was OK for a person to drive his car in the city center as long as he paid the surcharge, or would you still criticize him for not taking available public transportation?

• Let's say that a charge is added onto packaging materials which reflects the damage that such materials cause society—including both the costs of disposal and the aesthetic damage litter causes. If such a charge reflected all the damage, would you then feel that if a consumer wishes to buy the packaging and pay the extra costs, it's OK for him to do so, or would you still criticize such a consumer for being wasteful?

Sixty-seven percent of the environmentalists ($N = 12$) said they would still criticize the car driver; 40 percent ($N = 10$) would criticize the package buyer, though many respondents specifically stated that solid waste involved less of a moral issue than air and water pollution because of the absence of health effects. This attitude appeared repeatedly during deliberations in 1977 and 1978 within the interagency Resource Conservation Committee, set up under a provision of the 1976 Resource Conservation and Recovery Act to investigate the possibility of introducing a national product charge for packaging materials such as paper. Some studies done for the committee estimated that adding to the price of packaging material a charge reflecting its solid-waste cleanup costs would reduce consumption of such material only insignificantly. Most economists would react to this finding by stating that the addition of the charge would mean that packaging consumers were bearing the social cost of their behavior and that, given this, the quantity of packaging then consumed would be a matter of indifference. Yet the virtually universal reaction of those considering product charges was very different. The studies played an important role in diminishing enthusiasm for the product charge, because "they showed that product charges won't work." By "won't' work" was meant that they wouldn't reduce packaging consumption sufficiently. That "excessive" consumption of packaging was wrong was the initial assumption. A product charge would be accepted if it achieved a significant reduction in consumption; that it made packaging reflect its social cost, absent significant consumption effects, was regarded as unimpressive.

During the 1970 House hearings on air-pollution legislation, the following dialog took place between Representative David Satterfield and Dennis Hayes, the head of Environmental Action:

Table 14.7 Attitudes toward use of the word *criminal* and toward charges among environmentalists and Senate majority staffers ($N=20$).

Attitude toward use of word *criminal*	Attitude toward charges		
	For	Against	For experiments, but not full implementation
Favorable	25%	44%	19%
Unfavorable	50%	25%	25%

Satterfield I was interested to note when you started off that you mentioned something about this being a time we hear about law and order and then referring to "industrial criminals." I am interested in the use of this word. We hear more and more about "industrial criminals." Actually, you don't have any criminals in this country unless there is a law which is violated. Isn't that right?

Hayes Well, if you want to carry on a sort of semantic struggle, we are breaking a whole series of nature's laws.

Satterfield Don't you think you ought to state this is the type of criminality you are talking about rather than making a blanket statement that we have a lot of industrial criminals in this country?

Hayes Mr. Congressman, there are a whole series of things that are being done in industries that are fairly well documented right now which are contributing enormously to the degradation of the world, and probably in an irreversible manner. That kind of action, whether or not this body or a state legislature has seen fit to pass a law, is criminal. As I was using the term, a criminal is a person or institution who robs others of their rights to an ecologically balanced world. (*Air Pollution Control*, p. 642.)

Environmentalists and congressional staffers were asked a question designed to tap their attitudes toward using the morally condemnatory word *criminal* to talk about the behavior of polluters (See table 14.7). The question read:

Would you personally ever use the word "criminal" to describe the behavior of some individual or company which pollutes the environment? (If yes) Under what circumstances?

This question helped measure whether respondents took an attitude of moral condemnation toward polluters. The coders classified 76 percent of respondents ($N=21$) as having a basically favorable attitude towards the use of the word *criminal* to describe the behavior of many polluters. Furthermore, although the results should be interpreted with caution because of the small number of respondents who looked unfavorably on the use of the word and the small size of the universe, there was a definite tendency for those who reacted favorably to the use of *criminal* to be more hostile toward charges than those who reacted unfavorably.

Of the respondents to the preceding question who used the word *criminal* in a legal sense only and explicitly excluded a broader usage, 75 percent favored charges, whereas of the respondents who used it in a more inclusive sense (including polluting knowingly in general or making a broad general usage) not one favored charges ($N = 10$). There were similar results for the following question:

Do you think it is useful to use the word "blame" in reference to responsibility for pollution, or do you think that this is not a useful word?

Although environmentalists and majority staffers were more split on their attitude toward the word *blame* (58 percent coded as favorable), those who were favorable were more likely to oppose charges than those who were not.

Some statements on the use of the word *criminal* by environmentalists or majority staffers who oppose charges follow:

That's a problem with the economic approach, that it doesn't distinguish between willful and nonwillful. Economists don't see that distinction. These words are important in environmental policy. A crime against nature is a crime against society. I am part of a policy that has been adopted and that has an important goal. If I violate that policy, that's the same as if I rape, pillage, and burn. Society should be vengeful and punitive against violators of this policy. If my son dies of cancer, I want to blame somebody, and I want that somebody to be accountable.

I hold both the natural and the human environment on a pedestal. When a human knowingly is destroying the natural environment or the human environment of the rest of the public for his own economic gain, that is criminal.

I would use the word *criminal* on lots of occasions—much more often than it's generally used. It's easy to use it in cases like [that of the pesticide] Kepone. But I would also use it for any willful behavior— it's a crime against society to dump harmful materials in the air and water.

Breaking a pollution law is not comparable to breaking a traffic law. It is similar to crimes against persons.

It is criminal when there is a knowing abdication of responsibility.

These statements contrast with statements on the same subject by environmentalists and majority staffers who favored charges:

I don't use words like that. Polluters were responding to a market situation. You can't blame them. It's the way the system is structured. I wouldn't use the word *criminal*. I wouldn't even use the word *bad*. It's not a case of good actors or bad actors. It's a question of price signals.

The second strand in the concern of many Democratic staffers and environmentalists about charges as a "license to pollute" involved the equity-based worry that those who would pay the charge rather than reduce their polluting behavior would be those who could more easily afford to do so. Thus, not only would the use of charges rather than standards remove the social statement that pollution is wrong; it would also open the way to a situation where the rich continued to pollute, while the poor, who couldn't afford to pay the charge, would be the ones who would bear the burden of efforts to reduce pollution. Some statements:

There's always somebody who can pay the price and buy the right to pollute.

That's morally bad—it's a bad example for our children. We need other control than through money.

With a pollution charge you're telling people to whom that's pocket change to pass that money on to their customers. I don't think it's fair.

Say GE could afford the tax to pollute the Hudson. Do you want to grant someone the right to pollute and cause environmental and health consequences?

When the idea of using parking surcharges rather than direct parking restrictions to limit downtown driving was mentioned, one respondent stated: "My gut reaction is very strong on this. I hate it. I think it's elitist. I hate it for energy policy also." Specifically insofar as charges for air and water pollution are concerned, this view is based on an unambiguous error of reasoning. When people make decisions about whether to undertake some activity that gives them satisfaction but also has a cost, their decisions are indeed based not only on their preferences (the degree of satisfaction the activity gives them) and on the relative price of the activity, but also on their budget constraints. Increasing the price of activities that give people satisfaction will often produce a greater decrease in its consumption by the poor than by the rich. But polluting behavior rarely gives polluters any satisfaction over and above the costs saved by not having to install abatement devices. If polluters are charged for pollution, it is rational for any polluter, poor or rich, to reduce polluting behavior up to the point where the savings from reducing the pollution equal the charge. There is thus no equity issue involved. The only way it could be otherwise would be if polluters gained satisfaction from the very act of polluting—a "satisfaction," beyond the cost savings, that the poor might give up before the rich.

This is an error that some of those concerned about charges as a "license to pollute" made. However, although the equity argument does not properly apply to charges for air and water pollution, it does apply to some other issues involving the use of charges in environmental policy, such as increased tolls or parking surcharges (as opposed to parking bans) in transportation-control plans. Furthermore, it applies to many other issue areas, such as energy pricing, where economists have played an important role in policy debates. Democratic staffers and environmentalists can be expected to have heard such arguments there and applied their concern with equity issues in using pricing strategies to this area of environmental policy, where it clearly is inappropriate. Most economists do, furthermore, believe that individual public policies should not be designed with equity consideration in mind, and that equity issues should be addressed only through general cash transfer payments. (This lack of sympathy with the importance of equity issues in the design of public policy in specific areas may explain why economists defending charges have not bothered to take on the error in the view that pollution charges allow the rich to continue polluting while making the poor stop; I have seen no response to such contentions in the pro-charges literature by economists.)

The second element of the deeper belief system of which uneasiness about charges forms a part is a general uneasiness about the market. This came out most specifically when respondents stated as a disadvantage of charges the view coded under the category "One cannot set a monetary price on health or the environment." (This tied with "license to pollute" as the most frequently mentioned objection to using charges, cited by 37 percent of environmentalist respondents.)

Setting a charge places an explicit monetary price on pollution and, hence, on environmental quality. A number of respondents paled at that idea:

If you contaminate the environment, I don't see how you can pick a tax that is worthwhile.

We should make as a rule that rivers are treasures and not fair game.

I'm not really fond of the economists' tendency to be technocrats.

How do you figure out a tax for contaminating my wife's breast milk?

The unease was intuitive and rarely further justified. Yet it would appear to tap feelings that are widely shared, and not just by environmentalists. Some hint of the nature of the unarticulated thoughts came from the frequent observations that a charge approach would make pollution seem less a health issue and more a simple question of aesthetics. Other hints came from the view that a charge approach was more acceptable for solid waste than for air and water pollution. Unlike clean air and water, garbage (as well as recycled materials) are already in the price system.

That staffers on Capitol Hill saw the question of whether to use charges from a general political or ideological perspective is perhaps not so surprising; they are part of a political institution, and few had direct environmental experience. It was, however, a surprise to learn that many of the environmental lobbyists had a general political background rather than a specifically environmental one. Environmentalists were asked early in the interview what it was that initially got them interested in environmental problems. The question was asked to see whether there was any difference in support for charges between those whose initial interest was health-related and those whose interest was wilderness-related. It turned out, however, that 37 percent of the environmental lobbyists came into the environmental movement through general political concerns. An equally large group had their interest awakened by an interest in the outdoors, and 26 percent each initially had health-related interests and came to the job for non-values-related reasons ("just a job," "a good opportunity," etc.). Those who came to the environmental movement for broader political reasons were social activists of the 1960s who, like many of their generation, moved from civil rights to Vietnam to the environment. They were critical of many aspects of American society and saw environmental problems as examples of larger social problems.

Table 14.8 shows that the source of initial interest in the environment influenced dramatically the tendency of environmentalists and Senate Democratic staffers to favor charges. It also influenced the tendency to cite ideological arguments against charges (table 14.9).

Support for charges was nonetheless far greater among environmental lobbyists than among Democratic staffers. Two explanations may be cited: that environmentalists were more likely to be knowledgeable of the efficiency arguments for charges than were Democratic staffers, and

Table 14.8 Support for charges and source of environmental interest
(environmentalists and Senate Democratic staffers)($N=20$).

Source of environmental interest	Attitude toward charges		
	For	Against	For experiments, but not full implementation
Public interest	20%	80%	0
Health	40%	60%	0
"Just as a job" or personal reasons	50%	25%	25%
Outdoors, wildlife	17%	17%	66%

Table 14.9 Ideological disadvantages of charges cited and source of environmental
interest (environmentalists and Democratic staffers)($N=20$).

Source of environmental interest	"License to pollute"	"Can't set a price"
Public interest	60%	80%
Health	60%	20%
"Just as a job"	50%	25%
Outdoors, wildlife	16%	33%

that there appeared to be a greater tendency among environmentalists
than among Democratic staffers to feel ambiguous, rather than hostile,
about harnessing self-interest. A fair number of environmentalist respondents expressed a great respect for the power of self-interest to
motivate behavior at the same time that they feared exclusive reliance
on it. "Unfortunately, you can't turn everyone into an environmentalist
on principle," stated one environmentalist, sympathetic to charges,
who went on to describe himself as a "latent capitalist." After stating
a moral criticism of charges, another respondent added, "But I know
that's an idealistic approach to things, and I've been around." Another
expressed well the ambivalence many felt: "In the sixties, I was involved
in a Union Carbide case. I thought it was environmental violence. I
met a reporter from *Business Week*, who told me it was just a matter
of dollars and cents. At the time, I totally disagreed. Now I'm more
sympathetic. But, still, there is also a question of ethics and manners.
People don't keep clean just because of a knowledge of disease."

One might expect industry people to be sympathetic to charges because
they embody a market approach and because they encourage cost-
benefit thinking. But the trade-association people interviewed were, as
a group, the least informed about charge proposals. Although all stated
that they had heard such proposals made, they were less able to give

Table 14.10 Sources of new ideas for various respondents.

	Percentage saying they had no source	Mean no. of sources[a]
Industry	43	1.25
Environmentalist	25	1.67

	Percentage citing academics as source[a]	Percentage citing constituents as source[a]
Industry	7	57
Environmentalist	19	38

a. For those citing sources.

an account of arguments proponents made than were the other groups. (The industry people were able to cite a mean of 1.5 arguments proponents made for charges, compared with a mean of 1.8 arguments cited by environmentalists, 2.8 by Democratic staffers, and 3.0 by Republican staffers.) In many instances, the interviewers skipped asking the industry people many of the detailed questions about charges on the survey questionnaire because it appeared that some respondents knew so little about the issue that continued questioning would put them in the embarrassing position of having to answer "don't know" to a series of questions or to guess at or invent responses. In reply to the question whether any charge proponent had ever tried to convince them that charges were a good idea, 63 percent of environmentalists ($N=8$) and 100 percent of congressional staffers ($N=9$) said someone had spoken with them. Only 17 percent of industry respondents ($N=6$) said anybody had ever spoken with them.

Two differences between the industry and the environmental people help explain why the former were less knowledgeable about charges: They have fewer sources of information about policy alternatives, and they are more likely to limit their concerns about environmental policy to the issue of the stringency of EPA regulations.

Before the part of the interview that mentioned charges, respondents were given the statement "You sometimes hear that it is hard for people involved in the day-to-day formulation of policy to find the time to consider policy alternatives or policy approaches different from those which come up in day-to-day legislative or agency battles." They were then asked to mention their sources of new ideas. Table 14.10 compares the sources cited by the four groups inteviewed. Note the larger number of industry respondents citing no sources, the smaller mean number of sources cited by industry respondents. Note also the far greater

tendency of industry respondents to get information from their constituency (member firms) rather than other sources. (These sources,
presumably mostly environmental-affairs people with backgrounds in
science or engineering, are unlikely to direct the interest of trade-
association people toward new policy proposals.) Not only were industry
respondents unusually dependent for their information on member
firms, but they were unusually dependent on them for their views as
well. Both industry and environmentalist respondents were asked if
there were any group whose support for charges would make them
change their opinion. Sixty percent of industry respondents ($N=10$)
mentioned their constituency, a category not mentioned by a single
environmentalist.

Industry respondents seemed unusually single-minded in their focus
on the issue of the stringency of environmental regulations. For all the
groups interviewed, the question of what level of protection environmental regulation should require was a central one. But a fair number
of the industry respondents, during the portions of the interview dealing
with charges, kept trying to move the subject away from a discussion
of charges and back toward complaints that the EPA was imposing
unreasonable cleanup demands on industry.

Next to Democratic Senate staffers, industry respondents were the
most negative toward economic-incentive proposals. It was noted above
that industry opposition to charges would be rational, because the private
costs of charges to industry might often be greater than the cost of
standards. Hardly a single industry respondent had a knowledge of
charges sophisticated enough to see that point. As tables 14.11 and
14.12 indicate, the question of how much charges would cost industry
was nevertheless very much on respondents' minds. There were a number of generalized negative comments about pollution charges as "just
another tax," combined with fears that the charge would be used to
raise revenue (and thus raised periodically for that purpose). "Taxes
could go higher and higher until they became confiscatory," one respondent argued. "I can see this being ratcheted down, becoming a
never-ending moving target. The more you do, the more they ask you
to do." Another pointed out that nearby Montgomery County in sub-
urban Maryland had started using a tax on disposable bottles as a
revenue generator. Some of the industry respondents did have an inkling
of the fact that the private costs to them of charges might be greater
than those of standards. "Currently, you don't have to abate your
pollution below the standard, but an effluent tax is levied for every

Table 14.11   Arguments industry officials made for charges ($N=20$).

	Percentage of respondents who made argument
Charges make it in one's self-interest to reduce pollution	35
Internalize costs; "polluters should pay"	20
Encourage technical innovation	20
No advantages	15
Incentive to go beyond a standard	11
Promarket or antigovernment comment without further justification; prochoice	10
Those with very high costs or slow response times are given an option; put constant pressure on all	10
Know no arguments	10
Stated efficiency argument but did not understand it well	5
Less bureaucracy, more simplicity	5
Uniformity	5
"Based on cost-benefit considerations"	5

Table 14.12   Disadvantages of charges cited by industry respondents ($N=20$).

	Percentage of respondents who made argument
One should not have a choice of polluting, if willing to pay a charge; "license to pollute"	25
Dislike of money being spent as general revenues	25
Charges difficult to set; technical information needed; economic information needed	20
Charges inequitable (would not allow variances); would hit poor companies too hard; would concentrate industry	20
Government will set charges too high in order to raise revenue	15
Industry could just pass on costs and not control pollution	10
Industry uncertainty	10
Health damage should never be allowed; one cannot set a monetary value on health or the environment	5
Results are uncertain	5

point of pollution," one respondent said. "An effluent tax is in line with the zero-discharge concept. It's a zero-discharge regulation." Another pointed out that "the tax will continue even after correction of the pollution." A number of respondents expressed vague fears of a "double burden" of having to lower pollution levels and then have to pay a tax on residual pollution as well.

The small minority of industry respondents who favored charges did so for ideological reasons. Gary Knight, the chief lobbyist on environmental issues for the U.S. Chamber of Commerce, came (unlike most of the industry environmental lobbyists) out of a political background he was a staff member for a Republican representative and a legislative liaison to the Department of Housing and Urban Development during the Nixon administration. ("I loved the job because it gave me a chance to kick bureaucrats in the ass.") Knight said that he liked charges because "they're based on the economic method which most American businessmen understand, not on the hypocrisy that there's no threshold level below which there are no health effects or on the idea that we can get 95 percent abatement from everybody."

Political Strategy

By strategic reasons for opposing charges I mean those having to do with the effect of moving to charges on the process by which environmental issues are considered in the political system, the timing of their consideration, or the terms of the political debate. Strategic worries were that the effects of moving to charges would be unfavorable to environmentalists on those dimensions.

The responses by environmentalists and Democratic staffers showed that a significant part of the fear these groups had of charges was strategic in nature. Twenty-six percent of environmentalists stated that one disadvantage of charges was that it would be politically difficult to set them high enough; another 5 percent listed as a disadvantage the argument that reopening environmental legislation (something that would be required in order to grant the government authority to set pollution standards) would produce less stringent environmental laws, because the political climate was less favorable to the environment than when the laws were passed.

It should surprise nobody—but it is sometimes forgotten when discussing the difficulties of arousing interest in economic-incentive proposals—that the overriding concern of environmental and industry lobbyists is to work for adoption in environmental laws and regulations of what they consider appropriate requirements for levels of cleanup. Environmentalists spend most of their time fighting for stricter cleanup demands that would produce more improvements in environmental quality; industry spends most of its time fighting for more lenient cleanup demands that would produce smaller cost burdens. In a question asked early in the survey on criticisms of environmental policy, by far the most common criticism cited by both environmentalists and industry respondents was that environmental regulation called for inappropriate levels of protection. This criticism was cited by 84 percent of the environmentalist respondents ($N=19$) and 85 percent of the industry respondents ($N=20$)—far ahead of the next most common criticisms, which were never mentioned by more than a quarter of the respondents.

Environmentalists cited several reasons why a move toward economic incentives might produce a lowering of the level of cleanup demands. The first was simply that reopening environmental statutes to amendment would likely produce statutory language less favorable to environmentalists than the existing language. The second was the debate on the stringency of environmental demands that is phrased in terms of "What should the level of a charge be?" focuses attention on costs and thus hurts environmentalists, whereas debate that is phrased in terms of "What should a standard be?" focuses attention on health protection and thus helps environmentalists.

The current air and water environmental statutes were adopted in 1970 and 1972, when public support for the environmental movement was at its height and before a counteroffensive, aided by energy and inflation problems, could be organized. The statutes required standards to be set, and they did not authorize the establishment of charges. A move toward charges in the air and water areas would, therefore, require statutory changes. (The grounds for this fear may have declined with the increasing tendency to use the congressional reauthorization process, which occurs automatically and with regularity, to open statutes for revision. Indeed, the reauthorization of the Clean Air Act, about which Congress began to debate in 1981, certainly opened the statute to the possibility of revision. In a reinterview in 1981, one of those environmentalists most worried about the "reopening" issue stated that he still felt that environmentalists would want to minimize the amount of reopening; he was also afraid that it would be easier to hide a weakening of the statute under the guise of a move to charges.)

The second problem arises from the fact that it is common, in issues where there are tradeoffs among conflicting values, for political opponents to talk mostly about the importance of the value they are arguing for, without confronting the tradeoff between it and conflicting values head-on. Much political advocacy consists of pressing the case for the value of one goal while opponents press that for the value of the conflicting goal. Thus, one set of political actors mourns the devastation wrought by unemployment while another set paints the demoralizing results of inflation. This occurs for several reasons. Limitations of time and interest mean that most people are unable to consider most issues seriously, and that therefore the aspect of the issue they do hear about will often be important in influencing the choice they make. Furthermore, decisions involving clearly articulated conflicting goals are often psychologically painful. One's goal, if one wishes to promote a given choice, is therefore often to get the person to think about the positively regarded values realized by that choice and not to think about the negatively regarded ones the choice promotes. One is, in other words, concerned about the face the issue takes (Allison 1971, p. 168). Different faces produce different choices because the weights people attribute to positively regarded but conflicting values are not fixed but depend importantly on which of the values one is thinking about at the time of choice. Many decisions people make in their personal lives about things they buy would be far more difficult if objectives that conflicted with those the individual has in mind when making the purchase were articulated as clearly and forcefully as they often are in political debate. Decisions about buying chickens would be more difficult if pictures of the conditions under which chickens are commercially raised were repeatedly presented to people; decisions about traveling by air would be more difficult if consumers saw pictures of plane wreckage as frequently as they saw airline ads.

Both "saving health" and "saving money" are values that are widely regarded as positive. Environmentalists and industry representatives know this, and, seeking laws and regulations reflecting different levels of environmental protection as they do, they realize that they want to get people to think about the aspect of environmental regulation that best promotes the choice that they would hope to see people make. What some environmentalists were worried about is that, if environmental policy were set using charges, debates would be over the monetary amount of charge levels, and that this would inevitably make people think more about costs than otherwise. Debates on standards can be phrased—and are phrased by environmentalists—as a question of what is required to protect public health; this presents a face more favorable to environmentalists.

The analyst doubtless blanches at the approach of the policy advocate and wishes that people would confront the difficult tradeoff head-on. For an individual citizen, however, it might be more urgent that he be asked to confront head-on the tradeoffs in, say, food purchases, because they represent a larger expenditure than the individual's share of environmental protection. Analysts are, however, generally less quick to demand that people be forced to be confronted with tradeoffs in their decisions as consumers than in their decisions as citizens. This prejudice is justified, if at all, only for those actively formulating environmental policy; the total impact of their decisions (for society as a whole, not per person) is far greater.

One aspect of the strategic argument against charges is that making environmental legislation proceed using taxes would transfer committee jurisdiction in Congress from the committees where it currently lies to the tax-writing committees (Finance in the Senate, Ways and Means in the House). Accounts by advocates of economic incentives of the political difficulties charge proposals have encountered often refer to these jurisdictional consequences, but the implication is left that the source of opposition to such a shift is the self-interest of members of the committees who currently have power over such legislation and could be expected to oppose proposals that would limit their authority. Such an account of what is happening fosters the "throw up one's hands" approach to the political system that many academic economists tend to have anyway—a view that the rationality of economists' proposals flounders on the irrationality of the political system. If one places the efficiency advantages of charges against the paltriness of the desires of members of certain committees to retain personal power, the political system does indeed come out looking worthy of any contempt heaped upon it.

But things are not that simple. Committees have significant influence over the specific content of legislation passed by Congress. Committees can prevent bills from ever being considered on the floor. Members of Congress not on a committee frequently defer to the judgment of the committee on a bill. These features are rational in the American political system. In our system, in contrast with the situation in countries with parliamentary government and strong party discipline, the legislature gives bills genuine consideration rather than automatically passing every bill the executive sends. This genuine consideration requires that there exist some "gatekeeping" function so that Congress will not be swamped with bills and also some development of expertise in various areas of legislation. Specialization in turn would be useless unless there were some deference to the results of expert consideration. Gatekeeping and specialization are the two functions committees serve, and, in the context of a legislature that wishes to give bills genuine consideration, the unusual influence that committees have makes good sense.

Furthermore, it is not surprising that in different committees there should often be different dominant views on certain issues. Committees often tend to attract strong advocates (or, sometimes, strong opponents) of policy in a substantive area. Thus, the agriculture committees tend to attract farm representatives who favor programs to aid farmers. If different committees have different dominant viewpoints on an issue, and the committee out of which a bill comes has an important influence over the final content of the bill, then the jurisdiction of one committee or another over a bill may well change significantly a bill's final content. This is what environmentalists were worried about when they worried about the implications of moving from standards to charges for committee jurisdiction. The environmental committees were, especially at the time of our survey, filled with environmentalists—most prominently, at the time of the research, Senator Muskie and Representative Rogers. Neither of the chairmen of the tax-writing committees, by contrast, was particularly favorable to environmental issues. It was the fear that environmental legislation emerging from such committees would mandate lower degrees of environmental protection, not mere concerns over loss of personal power by members of existing committees, that dominated concerns about jurisdiction of environmentalists and their congressional allies.

The very first thing Karl Braithwaite, the Democratic staff director for the Senate committee, said in the course of his long interview was, "The first think I ask economists is 'Do you think environmental policy would be in better hands if it was under Russell Long?' " Probably part of the reason for his mentioning this so soon was to suggest that economists were impractical and didn't consider the "real-world" implications of their theories. But the concern about committee jurisdiction was clearly on Braithwaite's mind. Tariffs, he continued, were not intended as a revenue-raising measure, yet they went to the tax-writing committees. Braithwaite also went on to add that "Muskie hasn't approached the issue that way" and that it "isn't Leon's first concern."

Indeed, Leon Billings said the following in his interview: "The salient issue isn't jurisdiction. It's that effluent charges are a lousy idea. My experience on jurisdiction has always been that if you believe in something and are intelligent, you can do an end run."

When industry and environmentalist respondents were asked their reaction to a possible change in committee jurisdiction resulting from a move from standards to charges, both groups agreed that such a change would favor industry and hurt environmentalists. Of environmentalist respondents, 23 percent ($N=13$) stated they did not care one way or another if there were a shift in jurisdiction. Among those who did care, all reacted negatively to the idea of such a change. Of the industry respondents, 47 percent ($N=15$) did not care, but of those who did care, 63 percent ($N=8$) had a positive reaction and 37 percent a negative reaction (because the existing committees, they said, had developed expertise). Many of the industry people doubted whether any transfer of jurisdiction would take place. ("Muskie has enough seniority so that if you move the environment to Finance, Muskie will go to Finance.") But there was little opposition to the prospect, if it could be arranged. "We could probably get further in arguments on the merits with Russell Long than with Muskie and Rogers," one of the most important industry lobbyists told me. "When you walk in Public Works, all they care about is health." Responded another: "The business community would love it. It would be great to have Russell Long."

However, there was no tendency among the environmentalists who were worried about committee jurisdiction to be less sympathetic to charges than those who did not care. A cross-tabulation of attitudes on charges with attitude on committee jurisdiction showed no differences between the two groups of environmentalists. Similarly, possible industry support for charges because of hopes of getting more favorable committee consideration was inhibited by skepticism that such a jurisdictional transfer would take place, combined with a general failure to consider strategic implications of charges for industry, given a lack of thought given the issue in general. A number of industry respondents volunteered that they had never considered the impact of a move from standards to charges on committee jurisdiction before the question was asked them in our survey.

There was one more reason why environmentalists might be worried that a move to charges might produce a lower level of demands for environmental cleanup. In a situation where actors in the environmental policy process would have to spend time working for a shift from current regulatory approaches toward new ones, there may be elements of a Prisoner's Dilemma game. Imagine that industry and environmentalists might each value efficiency enough to devote 20 percent of their time working for greater efficiency in regulation (a goal on which both sides agree) while devoting 80 percent of their time to their other goals on which there is disagreement. Each side might be afraid that any time spent on working for a goal on which both sides agree would (unless the other side were willing to spend equal time working for the goal) be net time subtracted from working for the goals on which there is disagreement, and that this would increase the chances that one's opponent would prevail in those areas. Both sides might then fail to spend any time working for efficiency, and a less efficient result than desired would be produced.

Environmentalists were not as sympathetic to charges as were the Republican staffers, but they were the next-most-sympathetic group. Environmentalists and industry respondents were both asked which groups would be most likely to support charge proposals if they were introduced before Congress and which to oppose them. The dominant view among both groups was that industry would oppose such proposals and environmentalists would favor them—although there was less of a tendency to see environmentalists as supportive among environmentalists than among industry respondents. Environmentalists were also the most knowledgeable about charges.

The strategy for making environmentalists into supporters of charges must be to impress them strongly with the efficiency advantages of charges over standards. What one would then in effect be asking them to do in supporting charges would be to be willing to trade off the philosophical and strategic costs of going to charges against the efficiency advantages. Everybody is "for" efficiency in the sense that they would prefer achieving a given goal for less money than for more money, but whenever there are tradeoffs between efficiency and other goals the consensus over efficiency disappears. This is obvious in the huge number of political issues involving tradeoffs between efficiency and equity, but it also applies to moving from a standards-based to a charges-based approach to environmental regulation. If one believes that taxing rather than regulating pollution has various disadvantages along some dimension other than efficiency, then one must ask oneself how much (if any) of one's other goals one is willing to give up in order to get more efficiency. And even if one has no substantive objections to charges, one must still, if charges are not already in place, give up time that could otherwise be spent working for the attainment of other environmental goals to work for an efficient environmental policy. What economists must be—indeed, what they are when they work for charges—is lobbyists for efficiency, trying to influence people's preferences just the way other lobbyists do. They will have to compete against other lobbyists who are trying to get people to give other preferences a high weight. Such a view—that economists are joining the battle for preferences—goes against the grain of the self-image economists have of their profession. But it reflects the reality of what is happening and, perhaps, an appreciation for the nature of the political process.

Chapter 2

1. See Teitenberg 1974 and Teitenberg 1978 for discussions of theoretical advantages of geographic diversity. Harrison 1975 and Harrison 1978 consider the empirical advantages of geographic variation in automotive emission controls; the case study by Nichols in this volume evaluates the case of airborne benzene.

2. NEF values are calculated by aggregating aircraft operations by aircraft type and by flight path, estimating NEF values for each of the categories, and then aggregating to obtain the cumulative noise value. For a given aircraft class i and flight path j, the NEF is

$$\text{NEF}_{ij} = \text{EPNdB}_{ij} + 10 \log(d_{ij} + 16.67 n_{ij}) - 88$$

where EPNdB is the mean effective perceived noise level in decibels, d_{ij} is the number of daytime flights (7 A.M. to 10 P.M.), and n_{ij} is the number of nighttime flights. The value of NEF at a particular point on the ground is then

$$\text{NEF} = 10 \log \sum_{ij} \text{antilog}(\text{NEF}_{ij}/10).$$

3. The theoretical case for charges, standards, or marketable permits depends upon the shapes of the marginal-benefit and marginal-cost functions and the degree of uncertainty surrounding both functions. See Spence and Weitzman 1978.

Chapter 3

1. Because the Nelson study was completed before the 1979 modification in the compliance schedule, the costs and benefits presented here are for the retrofit rule as originally promulgated.

2. The results in table 3.6, based on Nelson 1978, contrast sharply with the results of the FAA study. Assuming a 10 percent discount rate, the FAA estimated total costs to be $440 million and total benefits to be $1.2 billion, for a net benefit of $760 million. The source of the difference in results is that the FAA's benefit estimate is almost four times the "high" benefit value Nelson employs. Although Nelson (1978) offers no specific explanation for the wide discrepancy in estimates, he does provide a persuasive critique of the FAA calculations (pp. 145–146).

3. Replacement is also an option to comply with the standard. The cost of replacing aircraft is difficult to calculate, as it depends upon the routes flown, the relative operating costs of the old and the new aircraft, the resale price of the old aircraft, and the purchase price and availability of the new aircraft.

4. The effect of charge strategies on the development of quieter planes is discussed below.

5. Setting percentage requirements for Part 36 aircraft operated at noisy airports is in the spirit of the current regulations, although other geographically varying regulations such as different maximum noise levels might also be appropriate.

6. The only questionable inclusions in the list of moderate-benefit airports are Dulles and Portland, which receive no benefits from the Part 36 extension. Their inclusion can be justified by the likelihood that future residential growth around Washington and

Portland will generate greater benefits than estimated by the DOT study, which used 1970 land-use information.

7. When standards are set for airport categories, the issue arises whether each airline would have to meet the percentage requirement for each airport or only for the airports as a group. Although requiring each airline to meet the standard at all airports would provide greater assurance that noise levels would be reduced at each airport, it would also increase the cost of control because eliminating non-Part 36 flights may be much more expensive at some airports than at others. On balance, it would probably be best to allow airlines to meet the timetable for the group of moderate-benefit airports.

8. Because a given aircraft may fly into all three types of airports, the percentage may be greater. It may also be smaller if the airlines partially meet the standards by rescheduling rather than retrofitting aircraft.

9. The airlines have complained that Part 36 values are "worst-case" estimates because they are based on maximum aircraft weight.

10. Because recent fuel-price increases have induced airlines to accelerate aircraft retirement, relaxing the requirements for older planes may no longer be necessary. There is no evidence, however, that Congress' different treatment of the first- and second-generation aircraft was due to such changes in retirement plans.

11. Rose-Ackerman 1973 shows that the response to a charge is not optimal when marginal benefits vary. In general, when the firm's desired output increases, a fixed charge allows too many emissions and a fixed standard too few.

12. The cost calculations are based on the simplifying assumptions that all JT3D and JT8D aircraft do not meet the Part 36 standards and that all JT9D aircraft do, and that all aircraft have a useful life of 10 years (after retrofit). Although more precise estimates of individual burdens would take into account the details of each airline's fleet (age composition, number of complying JT8D aircraft, and so forth), the cost impacts by airline would remain generally the same.

13. One potential difficulty with such a distribution scheme is its tendency to reduce market transactions and thereby limit the competitive determination of prices. The number of participants in a national aircraft-noise market would be so large, however, that such a "thinness" problem is not likely to arise. For a discussion of the possibility for an air-pollution market scheme for the Los Angeles basin, see Hahn and Noll 1982.

Chapter 4

1. Increasing landing fees through a noise charge is more consistent with having the authority or municipality underwrite the revenue bonds. Logan Airport is the major facility financed in this way. See Miller 1979 (pp. 85–98) for a discussion of airport pricing and financing. I thank Craig Miller for his major contribution to this chapter.

2. Airlines' noise-control efforts might not be optimal if options not generally available to airlines—relocation of residents, soundproofing of homes, and so forth—could achieve abatement more cheaply.

3. Enacting charges only at a single airport creates some efficiency complications. If a noise charge at Logan led to added traffic at Kennedy (where noise damages are even greater), overall efficiency might be reduced. See Lipsey and Lancaster 1957 for a general discussion of this "second best" problem.

Chapter 6

1. In December 1980, the EPA proposed two additional benzene regulations under section 112, covering ethylbenzene/styrene plants (45 *Fed. Reg.* 83448) and benzene storage vessels (45 *Fed. Reg.* 83952).

2. The original "final report" was issued in October 1977 (Mara and Lee 1977). In response to several serious criticisms, Mara and Lee revised their study and issued a new "final report" in May 1978. Errata sheets were issued in the fall of that year, correcting several errors. Unless otherwise indicated, the figures cited are drawn from the (corrected) 1978 version.

3. The exposure estimates do not cover the entire population exposed to some sources. In particular, the automobile exposures cover only standard metropolitan statistical areas with populations of 500,000 or more. Similarly, the exposure estimate for residents near service stations covers only urban areas. These restrictions probably have very little impact on the totals; Mara and Lee (1978, p. 94) estimated automobile-related exposures at less than 0.1 ppb for SMSAs of less than 500,000, and service station exposures, which are localized, are unlikely to be significant in nonurban areas.

4. See Nichols 1981b (p. 219) for a more detailed explanation.

5. This estimate is derived from table VIII-12 (corrected) of Mara and Lee 1978. They initially reported results that would yield an estimate of 6.5 million ppb-person-years, but those results incorporated an error in converting square miles to square kilometers. (See Nichols 1978.) The error was later corrected in errata sheets.

6. When initially printed, Mara and Lee's "final" report (1978) put the figure at 12.0 million ppb-person-years. That estimate, which reflected an error in their formula for calculating the area of concentric rings (see Nichols 1978), was later corrected in errata sheets.

7. See chapter 3 of Nichols 1981b for an extended discussion of the problems associated with subsidies.

8. See chapter 4 of Nichols 1981b for a summary of the arguments in favor of charges when marginal damage is constant.

9. See case study 3 for a discussion of the problems that arise when permit markets are not perfectly competitive.

10. See chapters 5–7 of Nichols 1981b for an analysis of the issues involved in choosing the "target" of regulation.

Chapter 7

1. For a useful discussion of the concept of thresholds in relation to carcinogenic and noncarcinogenic effects, see chapter II of National Academy of Sciences 1977.

2. The probability that no cancers will be observed in n animals if the risk is p is $(1-p)^n$.

3. The literature on these models is voluminous. For a useful review see Food Safety Council 1978, chapter 11.

4. This problem is well known in the literature. See, for example, Cornfield 1977, Hoel et al. 1975, and National Academy of Sciences 1977.

5. In its preliminary report, the CAG noted that this model was "expected to give an upper limit to the expected risk" (Albert et al. 1977, p. 1). The final report (Albert et al. 1979), however, contains no such qualifying statement. This change between the two

reports is typical of the way assumptions and uncertainties tend to get "lost" along the way.

6. It is not clear why the CAG chose the geometric mean rather than the more usual arithmetic one, but the choice makes little difference, given the other uncertainties involved, because the arithmetic mean is only 12.8 percent higher (0.027162 vs. 0.024074) than the geometric mean.

7. This approach faces several difficult obstacles, including how to elicit probability estimates from experts, how to combine the estimates of different experts with epidemiological and experimental data, and how to endow the process with sufficient credibility to withstand political and judicial scrutiny. Raiffa and Zeckhauser 1981 offer a useful but not conclusive discussion of these and other related issues.

8. Because the nonlinear models predict risks several orders of magnitude lower than the linear model at low doses, a weighted combination of dose-response functions will exhibit almost no curvature at low doses. The estimated linear coefficient under such a weighting scheme, however, will be lower than the linear model's. In particular, it will be about equal to the linear model's coefficient times the estimated probability that the linear model is the correct model.

9. Readers versed in applied welfare economics will recognize the two measures as compensating variation (CV) and equivalent variation (EV), which generally differ because of income effects. For small changes, however, where income effects are negligible, CV and EV measures converge. See Currie et al. 1971 for a general review of alternative measures of economic surplus.

10. For a useful and balanced discussion of discounting in this context, see section 4.4 of Raiffa et al. 1977.

11. Most cost-effectiveness analyses of life-saving programs ignore the time element, thus implicitly applying a zero discount rate. In their analysis of carcinogens in drinking water, Page et al. (1979) argue explicitly in favor of comparing steady-state costs and benefits (which yields essentially the same result as a zero discount rate) on the grounds of intergenerational equity.

12. This is the annual payment that, when discounted at 5 percent over 35 years, yields a present value of $1 million:

$$\sum_{t=0}^{34} \frac{58,164}{1.05^t} = 1 \times 10^6.$$

13. The calculation is straightforward:

$$\sum_{t=15}^{34} \frac{58,164}{1.05^t} = 366.100.$$

Chapter 8

1. The EPA estimates that emissions could be monitored continuously by a gas chromatograph at a total annualized cost of about $9,000 per plant—roughly 2 percent of the estimated control costs for meeting the proposed standard (U.S. Environmental Protection Agency 1980, p. 5-21).

2. The assumed reduction in conversion rates with controls increases emissions by a factor of $0.10/0.055 = 1.82$. Thus, "97 percent" control reduces emissions to $0.03 \times 1.82 = 0.0545$ of their original level, a net reduction of 94.55 percent. The calculation for "99 percent" control is similar: $0.01 \times 1.82 = 0.0182$, a net reduction of 98.18 percent.

3. The basic technical data are given in Lawson 1978. The estimates used in U.S. Environmental Protection Agency 1980 are adjusted to reflect more recent prices.

4. See table 5-4 of U.S. Environmental Protection Agency 1980 for a comparison of the parameter values used in the two estimates.

5. For carbon adsorption at 97 percent, net annual costs are about 4 percent higher at 56 percent capacity utilization than at full capacity. With thermal incineration, costs fall with capacity utilization, though not in proportion (U.S. Environmental Protection Agency 1980, pp. 5-21–5-24).

6. Plant-specific exposure factors were not reported in U.S. Environmental Protection Agency 1980. They can be calculated, however, with data in appendix E of that document.

7. See Miller 1978 for a "state-of-the-art" review of techniques for estimating exposures.

8. The 90 percent standard considered here is a *net* 90 percent reduction; the conversion rate is held constant or the nominal level of control is increased sufficiently to overcome a reduction in the conversion rate.

9. With a V that high, however, 99 percent would yield greater net benefits; the marginal-cost-effectiveness ratio of shifting from 90 percent to 99 percent is $31.8 million.

10. Information on the potential conversion to *n*-butane and on the EPA's efforts to confirm Monsanto's plans was obtained from Deborah Taylor of the EPA in late 1979.

11. If costs were cut in half and only Monsanto controlled, estimated net benefits would rise by $560,000/2 = $280,000, from $437,000 to $717,000. With costs cut in half, the estimated net benefits associated with controlling Tenneco at 99 percent for $V = $3 million are ($3 \times 10^6)(0.05131) - 0.5(224,000) = $42,000. Thus, with costs cut in half and both plants controlled, estimated net benefits rise to $717,000 + $42,000 = $759,000.

12. All of the testimony by CMA witnesses cited below was presented orally at the EPA hearings held on August 21, 1980, in Washington, D.C. The figures cited are drawn from the written version of that testimony, "CMA Presentation at the EPA Hearing on the Proposed NESHAP for Benzene Emissions from Maleic Anhydride Plants."

13. With an annual capacity of 10.7×10^6 kg maleic anhydride per year, the Pfizer plant is smaller than any of those analyzed by the EPA. Costs were estimated using a linear extrapolation downward from the next two smallest plants for which 97 percent control costs were estimated, Tenneco and Reichold (Illinois). That yields an estimated annual cost of $199,000. The estimated annual reduction in emissions under the proposed standard is 678,000 kg benzene, based on capacity, an uncontrolled emission rate of 0.067 kg benzene per kg maleic anhydride, and an effective control rate of 94.5 percent. The estimated annual reduction in exposure is 11,119 ppb-person-years, based on concentrations from EPA's model-plant dispersion modeling (U.S. Environmental Protection Agency 1980, p. E-12) scaled down to the smaller plant size and on population estimates contained in the testimony of CMA witnesses Galluzzo and Glassman (1980). The estimated annual reduction in fatalities is 0.0038, based on the reduction in exposure and the CAG risk factor of 0.339×10^{-6} fatalities per ppb-person-year.

14. The CMA did not prepare estimates for Koppers or for Reichold (New Jersey), because those plants have closed. The "EPA estimate" for Pfizer was computed by the author from EPA dispersion-modeling results and CMA population figures, as discussed in note 13. The "CMA estimates" of exposure factors were calculated by dividing estimated exposures by estimated emissions, as reported in Galluzzo and Glassman 1980.

Chapter 9

1. The tighter standards, however, are imposed by the state of California, not by federal authorities.

2. For a more general analysis of the efficiency advantages of exposure charges, see chapter 6 of Nichols 1981b.

3. Strictly, the optimal charge is independent of control costs only if marginal damage is invariant with exposure. As discussed in chapter 7, the expected marginal risk for benzene and other carcinogens will be constant even if one believes that there is only a small probability that the linear model is correct.

Chapter 10

1. Clean Air Act Amendments of 1977, P.L. 95-95, 91 Stat. 685 (codified at 42 U.S.C. § 7401–7642).

2. CAAA of 1977, Sec. 109, 42 U.S.C. § 7409.

3. CAAA of 1977, Sec. 111–112, 42 U.S.C. § 7444.

4. CAAA of 1977, Title II, Sec. 202, 42 U.S.C. § 7521.

5. CAAA of 1977, Title II, Sec. 202 (a) (3) (A) (ii), 42 U.S.C. § 7521 (a) (3) (A) (ii).

6. CAAA of 1977, Sec. 110, 42 U.S.C. § 7410.

7. 41FR 55525, Dec. 21, 1976.

8. CAAA of 1977, Sec. 172, 42 U.S.C. § 7501–7508.

9. U.S. EPA, O.P.E., Emission Reduction Banking and Trading Project, "Planning for Clean Air and a Strong Economy," Washington, D.C., July 1979.

10. CAAA of 1977, Sec. 160–169, 42 U.S.C. § 7470–7479.

11. Senate Report 91-1196, 91st Congress, 2nd session, p. 2 (1970).

12. *Sierra Club et al.* v. *Ruckelshaus*, 344 F. Supp. 253 (D.D.C.).

13. 38FR 18986, July 16, 1973.

14. 39FR 42510, Dec. 5, 1974.

15. See note 10 *supra*.

16. *Alabama Power* v. *Costle*, 606 F. 2nd 1068 D.c.C., 1979.

17. 45FR 52676, August 7, 1980.

18. CAAA of 1977, Sec. 169A. 42 U.S.C. § 7491.

19. 45FR 34762, May 22, 1980.

20. 43FR 26388, June 19, 1978.

21. 43FR 26401, June 19, 1978.

Chapter 11

1. 38FR 18987, July 16, 1973.

2. 38FR 18988, July 16, 1973.

3. CAAA of 1977, Sec. 169b, 42 U.S.C. § 7491.

4. CAAA of 1977, Sec. 126, 42 U.S.C. § 7426.

5. CAAA of 1977, Sec. 169 (3), 42 U.S.C. § 7479.

6. U.S. EPA Office of Air Quality Planning and Standards, "Guidelines for Determining Best Available Control Technology," Research Triangle Park, North Carolina, June 14, 1978.

7. Minneapolis–St. Paul has been experimenting since February 1979 with a system of emission density zoning as a means of maintaining ambient standards, with the support of Federal Air Quality Technical Assistance grants. See Jerry A. Kurtzweg and Cristina J. Nelson, "Clean air and economic development: An urban initiative," *JAPCA* XXX (November 1980): 1187–1193.

8. CAAA of 1977, Sec. 120; 42 U.S.C. § 7420.

Bibliographies

Case Study 1

Air Transport Association. ATA Data Book. Washington, D.C., 1978.

Alexandre, A., and J. P. Barde. Aircraft noise charges. *Noise Control Engineering* (September–October 1974).

Anderson, Robert, Jr., Robert O. Reid, and Eugene P. Seskin. An Analysis of Alternative Policies for Attainment and Maintenance of a Short-term NO_2 Standard. Mathtech, Inc., Princeton, N.J., 1979.

Baumol, William J., and Wallace E. Oates. *The Theory of Environmental Policy.* Englewood Cliffs, N.J.: Prentice-Hall, 1975.

Baumol, William J., and Wallace E. Oates. *Economics, Environmental Policy, and the Quality of Life.* Englewood Cliffs, N.J.: Prentice-Hall, 1979.

Baxter, William F., and Lillian R. Altree. Legal aspects of airport noise. *Journal of Law and Economics* 15 (1972): 99–102.

Blitch, Stephen G. Airport noise and intergovernmental conflict: A case study in land use parochialism. *Ecology Law Quarterly* 5 (1976): 669.

Carlin, A., and R. E. Park. Marginal cost pricing of airport runway capacity. *American Economic Review* (1970): 310–319.

Civil Aeronautics Board. Airport Activity Statistics of Certified Routes Air Carriers, 12 Months Ending December 31, 1976. Washington, D.C.

Civil Aeronautics Board. Letter of P. J. Bakes to Mr. Richard Heath, Director of Airports, San Francisco Airports Commission, February 13, 1979.

Coase, Richard. The problem of social cost. *Journal of Law and Economics* 3 (October 1960): 1–44.

Council on Wage and Price Stability. Comments of the Council on Wage and Price Stability: The Noise Charge Approach to Reducing Airport Noise. FAA docket 16279, 1977.

Dales, J. H. *Pollution, Property, and Prices.* University of Toronto Press, 1968.

David, Martin, et al. Marketable Effluent Permits for the Control of Phosphorous Effluent in Lake Michigan. Social Systems Research Institute working paper, University of Wisconsin, Madison, 1977.

Drayton, William, Jr. Comment. In Ann F. Friedlaender (ed.), *Approaches to Controlling Air Pollution.* Cambridge, Mass: MIT Press, 1978.

Federal Aviation Administration. FAA Part 36 Compliance Regulation: Final Environmental Impact Statement. Washington, D.C., 1976.

Federal Aviation Administration. Impact of Noise on People. Washington, D.C.: U.S. Department of Transportation, 1977.

Federal Aviation Administration. Aircraft Noise Monitoring Report: Washington National Airport, Dulles International Airport, monthly reports for October 1978 and November 1978. Washington, D.C.: U.S. Department of Transportation, 1978.

Federal Aviation Administration. Forecasts of Aviation Activity, 1980–2000. Washington, D.C., 1979.

Hahn, Robert W., and Roger G. Noll. Designing a market for tradable emissions permits. In Wesley A. Magat (ed.), *Reform of Environmental Regulation*. Cambridge, Mass.: Ballinger, 1982.

Harrison, David, Jr. *Who Pays for Clean Air?* Cambridge, Mass.: Ballinger, 1975.

Harrison, David, Jr. Controlling automotive emissions: How to save more than $1 billion per year and help the poor too. *Public Policy* 25, no. 4 (Fall 1977): 527–553.

Harrison, David, Jr., and Paul R. Portney. Who loses from reform of environmental regulation. In Wesley A. Magat (ed.), *Reform of Environmental Regulation*. Cambridge, Mass.: Ballinger, 1982.

Harrison, David, Jr., and Daniel L. Rubinfeld. Hedonic housing prices and the demand for clean air. *Journal of Environmental Economics and Management* 4 (1978a): 110–126.

Harrison, David, Jr., and Daniel L. Rubinfeld. The distribution of benefits from improvements in urban air quality. *Journal of Environmental Economics and Management* 5 (1978b): 313–332.

Hastings, E. C., A. W. Mueller, and J. R. Hamilton. Noise Data for Twin-Engine Commercial Jet Aircraft Flying Conventional, Steep, and Two-Segment Approaches. NASA technical note D-8441, Washington, D.C., 1977.

Jacoby, Henry, and John Steinbruner. *Cleaning the Air*. Cambridge, Mass.: Ballinger, 1973.

Jaynes, J. E. Analysis of Take-off Noise Data at Logan International Airport. S.M. thesis, Massachusetts Institute of Technology, 1978.

Kneese, Allen V., and Charles L. Schultze. *Pollution, Prices, and Public Policy*. Washington, D.C.: Brookings Institution, 1975.

Kozicharow, Eugene. Flight procedures to curb noise studied. *Aviation Week and Space Technology*, April 12, 1976.

Leone, Robert A., and John E. Jackson. The Political Economy of Federal Regulatory Activity. Working paper 78-6, Harvard University Graduate School of Business Administration, 1978.

Lipsey, R. G., and Kelvin Lancaster. The general theory of the second best. *Review of Economic Studies* 24 (1957): 11–32.

Little, I. M. D., and K. M. McLeod. The new pricing policy of the British Airports Authority. *Journal of Transport Economics and Policy* 6, no. 2 (May 1972): 101–115.

Massachusetts Port Authority. The Effects of Limiting Night Flights at Logan Airport. Paper presented to the Advisory Committee on Night Noise Limitations, 1976.

Massport Master Plan Study Team. Draft Logan Airport Master Plan Study. Report prepared for Massachusetts Port Authority, 1975.

Meindle, H. G., L. S. Sutherland, H. Spiro, C. Bartel, and D. Pres. Costs and National Noise Impact of Feasible Solution Sets for Reduction of Airport Noise. Wyle research report WR 75-9, El Segundo, California, 1976.

Meyer, John R., et al. *Airline Deregulation: The Early Experience*. Cambridge, Mass.: Auburn House, 1981.

Milch, Jerome E. Feasible and prudent alternatives: Airport development in the age of public protest. *Public Policy* 24 (winter 1976): 81–109.

Miller, Craig L. The Use of Economic Incentives to Encourage Noise Abatement: The Airport Operators' Perspective. M.S. thesis, Massachusetts Institute of Technology, 1979.

Mills, Edwin S., and Lawrence J. White. Government policies toward automotive emissions control. In Ann F. Friedlaender (ed.), *Approaches to Controlling Air Pollution*. Cambridge, Mass.: MIT Press, 1978.

Motor Vehicles Manufacturers Association. Motor Vehicles Facts and Figures. Detroit, 1980.

Muskin, J. B., and J. A. Sorrentino, Jr. Externalities in a regulated industry: The aircraft noise problem. *American Economic Review* 67, no. 4 (September 1977): 770–774.

National Academy of Sciences. Noise Abatement: Policy Alternatives for Transportation. Washington, D.C., 1977.

Nelson, Jon. *Economic Analysis of Transportation Noise Abatement*. Cambridge, Mass.: Ballinger, 1978.

Nelson, Jon. *Airports and Property Values: A Survey of Recent Evidence*. University Park: Pennsylvania State University Press, 1979.

Nierenberg, Roy A. Incentives versus regulation: The case for airport noise charges. *George Mason University Law Review* 2 (1978): 167–208.

Olsen, Mansor, and Richard Zeckhauser. The efficient production of external economics. *American Economic Review* 60 (June 1970): 512–517.

Palmer, Adele R., et al. Economic Implications of Regulating Chlorofluorocarbon Emissions from Nonaerosol Applications. Report R-2524-EPA, Rand Corp., Santa Monica, Calif., 1980.

Pearce, David. *Charging for Noise*. Paris: Organization for Economic Cooperation and Development, 1976.

Plott, C. R., and V. L. Smith. An experimental examination of two exchange institutions. *Review of Economics and Statistics* 45 (February 1978): 133–153.

Roberts, Marc J., and Michael Spence. Effluent charges and licenses under uncertainty. *Journal of Public Economics* 5 (1976): 193–208.

Rose-Ackerman, Susan. Effluent charges: A critique. *Canadian Journal of Economics* 6, no. 4 (1973): 260–267.

Rose-Ackerman, Susan. Market models for pollution control: Their strengths and weaknesses. *Public Policy* 25, no. 3 (summer 1977).

Russell, Clifford S. What can we get from effluent charges? *Policy Analysis* 5, no. 2 (spring 1979).

Simpson, R. W. Ranking Airline Noise Performance—A Review of the Demonstration Program at Logan Airport. Cambridge, Mass.: Flight Transportation Associates, 1979.

Spence, A. Michael, and Martin L. Weitzman. Regulatory strategies for pollution control. In Ann F. Friedlaender (ed.), *Approaches to Controlling Air Pollution*. Cambridge, Mass.: MIT Press, 1978.

Straszheim, Mahlon R. *The International Airline Industry*. Washington, D.C.: Brookings Institution, 1969.

Straszheim, Mahlon R. Efficiency and equity considerations in the financing of noise abatement activities at airports. *International Journal of Transportation Economics*. 2 (April 1975): 3–15.

Suurland, Jan A. *Noise Charges in the Netherlands*. Paris: Organization for Economic Cooperation and Development, 1977.

Teitenberg, T. M. Derived decision rules for pollution control in a general equilibrium space economy. *Journal of Environmental Economics and Management* 1, no. 1 (May 1974): 3–16.

Teitenberg, T. M. Spatially differentiated air pollution emission charges: An economic and legal analysis. *Land Economics* 54, no. 3 (August 1978): 265–277.

Terkla, David G. The Revenue Capacity of Effluent Charges. Doctoral diss., University of California, Berkeley, 1980.

Urban Systems Research and Engineering. Land Use Control Strategies for Airport Impacted Areas. Washington, D.C.: U.S. Department of Transportation, 1972.

U.S. Congress, House of Representatives, Committee on Public Works and Transportation, Subcommittee on Aviation. Current and Proposed Federal Policy on Abatement of Aircraft Noise. Serial no. 94-36, 94th Congress, 1st and 2nd sessions. Washington, D.C.: Government Printing Office, 1976.

U.S. Congress, House of Representatives, Committee on Public Works and Transportation, Subcommittee on Aviation. Airport and Aircraft Noise Reduction. 95th Congress, 1st session. Washington, D.C.: Government Printing Office, 1977.

U.S. Department of Housing and Urban Development. Aircraft Noise Impact. Washington, D.C., 1972.

U.S. Department of Transportation. Airport Noise Reduction Forecast (Summary Report for 23 Airports). Springfield, Va.: National Technical Information Service, 1974.

U.S. Department of Transportation. Aviation Noise Abatement Policy. Washington, D.C.: Federal Aviation Administration, 1976.

U.S. Department of Transportation. FAA Aviation Forecasts Fiscal Years 1979–1990. Washington, D.C.: U.S. Department of Transportation, Federal Aviation Administration, Office of Aviation Policy, 1978.

U.S. Environmental Protection Agency. Report to the Congress on Noise. Washington, D.C.: Government Printing Office, 1972.

U.S. Environmental Protection Agency. Report of Operations Analysis Including Monitoring, Enforcement, Safety and Costs, NTID report 73.3. Springfield, Va.: National Technical Information Service, 1973.

U.S. Environmental Protection Agency. The Effects of Noise. Washington, D.C., 1974.

U.S. Environmental Protection Agency. Aviation Noise: The Next Twenty Years. Report EPA 55019-80-319. Washington, D.C.: U.S. Environmental Protection Agency Office of Noise Abatement and Control, 1980.

Weitzman, Martin. Prices vs. quantities. *Review of Economic Studies* (October 1974).

Zeckhauser, Richard, and Anthony Fisher. Averting Behavior and External Diseconomics. Harvard University, John F. Kennedy School of Government, discussion paper 41D, April 1976.

Case Study 2

Acton, J. Evaluating Public Programs to Save Lives. Rand Corporation report R-950-RC. Santa Monica, Calif., 1973.

Aksoy, M., et al. Leukemia in shoe workers exposed chronically to benzene. *Blood*, 44 (1974): 837–841.

Aksoy, M. Types of leukemia in chronic benzene poisoning: A study in thirty-four parts. *Acta Haematologica* 55 (1976): 65–72.

Aksoy, M. Testimony before Occupational Safety and Health Administration, Washington, D.C., July 13, 1977.

Albert, R., et al. Carcinogen Assessment Group's Preliminary Report on Population Risk to Ambient Benzene Exposures. U.S. Environmental Protection Agency, 1977.

Albert, R., et al. Carcinogen Assessment Group's Final Report on Population Risk to Ambient Benzene Exposures. U.S. Environmental Protection Agency, 1979.

Arthur D. Little and Co., Inc., 1977. Technology Assessment and Economic Impact Study of an OSHA Regulation for Benzene. Report prepared for OSHA, 1977.

Bailey, M. J. *Reducing Risks to Life*. Washington, D.C.: American Enterprise Institute, 1980.

Blomquist, G. Valuation of Life: Implications of Automobile Seat Belt Use. Ph.D. diss., University of Chicago, 1977.

Calkins, D. R., et al. Identification, Characterization, and Control of Potential Human Carcinogens: A Framework for Federal Decision-making. Staff paper, Office of Science and Technology Policy, Washington, D.C., 1979.

Chemical Manufacturers' Association. Cost considerations. In CMA presentation at the EPA hearing on the proposed NESHAP for benzene emissions from maleic anhydride plants, Washington, D.C., 1980.

Cornfield, J. Carcinogenic risk assessment. *Science* 198 (1977): 693–699.

Crump, K. S., et al. Fundamental carcinogenic processes and their implications for low dose risk assessment. *Cancer Research* 36 (1976): 2973–2979.

Currie, J. M., J. A. Murphy, and A. Schmitz. The concept of economic surplus and its use in economic analysis. *Economic Journal* 81 (1971): 741–799.

Dillingham, A. E. The Injury Risk Structure of Occupations and Wages. Ph.D. diss., Cornell University, 1979.

Food Safety Council. A System for Food Safety Assessment. Final report of the Scientific Committee, Columbia, Maryland, 1978.

Galluzzo, N. B., and D. Glassman. Recalculated benzene emissions, population exposures, and risk. In CMA Presentation at the EPA hearing on the proposed NESHAP for benzene emissions for maleic anhydride plants, Washington, D.C., 1980.

Harrison, D. *Who Pays for Clean Air?* Cambridge, Mass.: Ballinger, 1975.

Hoel, D. G., et al. Estimation of risks of irreversible, delayed toxicity. *Journal of Toxicology and Environmental Health* 1 (1975): 133–151.

Infante, P. F., et al. Leukemia in benzene workers. *Lancet* 2 (1977): 76–78.

Jandl, J. H. A Critique of EPA's Assessment of Health Effects Associated with Atmospheric Exposure to Benzene. Submitted by American Petroleum Institute to U.S. EPA, 1977.

Lamm, S. H. Oral presentation to the EPA for the American Petroleum Institute, Washington, D.C., 1980.

Lawson, J. F. Maleic Anhydride — Product Report. Report prepared by Hydroscience, Inc., for U.S. EPA under contract 68-02-2577, 1978.

Luken, R., and S. Miller. Regulating Benzene: A Case Study. U.S. Environmental Protection Agency, 1979.

Mantel, N., and W. R. Bryan. "Safety" testing of carcinogenic agents. *Journal of the National Cancer Institute* 27 (1961): 455–470.

Mara, S. J., and S. S. Lee. Human Exposures to Atmospheric Benzene. Report prepared by Stanford Research Institute for U.S. Environmental Protection Agency, 1977.

Mara, S. J., and S. S. Lee. Assessment of Human Exposure to Atmospheric Benzene. Report prepared by SRI International for U.S. Environmental Protection Agency, 1978.

Maugh, T. H. Chemical carcinogens: How dangerous are low doses? *Science* 202 (1978): 37–41.

Miller, C. Exposure Assessment Modelling: A State-of-the-Art Review. Report EPA-600/3-78-065 prepared for U.S. Environmental Protection Agency.

National Academy of Sciences. Health Effects of Benzene: A Review. Report by Committee on Toxicology, Washington, D.C., 1976.

National Academy of Sciences. Drinking Water and Health. Report by Safe Drinking Water Committee, Washington, D.C., 1977.

Nichols, A. L. Errors Noted in "Assessment of Human Exposures to Atmospheric Benzene." Unpublished paper, 1978.

Nichols, A. L. Alternative strategies for regulating airborne benzene. In D. Harrison et al., Incentive Arrangements for Environmental Protection, report to U.S. Environmental Protection Agency under grant R805446010, 1981a.

Nichols, A. L. Choosing Regulatory Targets and Instruments, with Applications to Benzene. Ph.D. diss., Harvard University, 1981b.

Ott, M. G., et al. Mortality Among Individuals Occupationally Exposed to Benzene. Exhibit 154, OSHA benzene hearings, 1977.

Page, T., R. Harris, and J. Bruser. Removal of Carcinogens from Drinking Water: A Cost-Benefit Analysis. Social science working paper 230, California Institute of Technology, 1979.

PEDCo Environmental, Inc. Atmospheric Benzene Emissions. Report EPA-450/3-77-029 submitted to U.S. Environmental Protection Agency, 1977.

Raiffa, H., W. B. Schwartz, and M. C. Weinstein. Evaluating health effects of societal decisions and programs. In *Decision Making in the Environmental Protection Agency*, volume IIB. Washington, D.C.: National Academy of Sciences, 1977.

Raiffa, H., and R. Zeckhauser. Reporting of Uncertainties in Risk Analysis. Unpublished paper, 1981.

Redmond, C. K., et al. Cancer experience among coke by-product workers. *Annals of the New York Academy of Sciences* 271 (1976): 102–115.

Regulatory Analysis Review Group. U.S. Environmental Protection Agency's Proposed Policy and Procedures for Identifying, Assessing and Regulating Airborne Substances Posing a Risk of Cancer. Report submitted to Council on Wage and Price Stability, 1980.

Schelling, T. C. The life you save may be your own. In *Problems in Public Expenditure Analysis*, S. Chase, ed. Washington, D.C.: Brookings Institution, 1968.

Schneiderman, M. A., et al. From mouse to man. *Annals of the New York Academy of Sciences* 246 (1975): 237–248.

Smith, R. S. *The Occupational Safety and Health Act*. Washington, D.C.: American Enterprise Institute, 1976.

Thaler, R., and S. Rosen. The value of saving a life. In *Household Production and Consumption*, N. Terleckyj, ed. New York: National Bureau of Economic Research, 1976.

Thorpe, J. J. Epidemiologic survey of leukemia in persons potentially exposed to benzene. *Journal of Occupational Medicine* 16 (1974): 375–382.

Turner, F. C., et al. Cost of Benzene Reduction in Gasoline to the Petroleum Refining Industry. Report EPA-450/2-78-021 prepared for U.S. Environmental Protection Agency, 1978.

Tversky, A., and D. Kahneman. Judgment under uncertainty: Heuristics and biases. *Science* 185 (1974): 1124–1131.

U.S. Environmental Protection Agency. Assessment of Health Effects of Benzene Germane to Low-Level Exposure. Office of Research and Development report EPA-600/1-78-061, 1978a.

U.S. Environmental Protection Agency. Standard Support Environmental Impact Statement for Control of Benzene From the Gasoline Marketing Industry. Draft report, Office of Air Quality Planning and Standards, Research Triangle Park, N.C., 1978b.

U.S. Environmental Protection Agency. Benzene Emissions from the Ethylbenzene/Styrene Industry. Emission Standards and Engineering Division report EPA-450/3-79-035a, 1979.

U.S. Environmental Protection Agency. Benzene Emissions from Maleic Anhydride Industry—Background Information for Proposed Standards. Emission Standards and Engineering Division report EPA-450/3-80-001a, 1980.

Viscusi, W. K. Labor market valuations of life and limb: Empirical evidence and policy implications. *Public Policy* 26 (1978): 359–386.

Zapp, J. A. Testimony at OSHA benzene hearings, Washington, D.C., docket H-059, 1977.

Zeckhauser, R. Procedures for valuing lives. *Public Policy* 23 (1975): 419–464.

Zeckhauser, R., and D. Shepard. Where now for saving lives? *Law and Contemporary Problems* 40 (1976): 5–45.

Case Study 3

Ackerman, Bruce, and William Haskell. Beyond the New Deal: Coal and the Clean Air Act. *Yale Law Journal* 89 (1980): 1466–1571.

Ahmed, A. Karim, and Frederica P. Perera. Respirable Particles. New York: National Resources Defense Council, 1978.

APCA (Air Pollution Control Association). Atmospheric Dispersion Modeling (reprint). Pittsburgh, 1980.

Atkinson, Scott, and Donald Lewis. Determination and implementation of the optimal air quality standards. *Journal of Environmental Economics and Management* III (1976): 363–380.

Bingham, Taylor, et al. Allocative and Distributive Effects of Alternative Air Quality Attainment Policies. Research Triangle Institute, Research Triangle Park, N.C., 1974.

Business Roundtable, Air Quality Project. The Effects of Prevention of Significant Deterioration on Industrial Development. Prepared by Arthur D. Little, Inc., Cambridge, Mass., 1980.

Caves, Richard E. Market structure and embodied technical change. In Richard E. Caves and Marc Roberts, *Regulating the Product*. Cambridge, Mass.: Ballinger, 1975.

Dames and Moore, Inc. An Investigation of Prevention of Significant Deterioration (PSD) of Air Quality and Emission Offset Permitting Process. National Commission of Air Quality, Washington, D.C., 1980.

de Nevers, Noel. Some alternative PSD policies. *Journal of the Air Pollution Control Association* XXIX (November 1979).

ERT (Environmental Research and Technology, Inc.). Impact of the EPA Class II Significant Deterioration Regulations on Multiple Industries: Size and Siting Limitations. Prepared for Federal Energy Administration, Washington, D.C., 1977.

Ferris, Benjamin G. Health effects of exposure to low levels of regulated air pollutants. *JAPCA* 28 (1978): 482–497.

Freeman, A. Myrich. *The Benefits of Environmental Improvement; Theory and Practice.* Baltimore: Johns Hopkins University Press, 1979.

GCA. Acid Rain Information Book. Prepared for U.S. Department of Energy, Office of Environment, Bedford, Mass., 1980.

Galloway, J. N., and D. M. Whelpdale. An atmospheric sulfur budget for North America. *Atmospheric Environment* XIV (1980): 409-417.

Grad, Frank, et al. *The Automobile and the Regulation of Its Impact on the Environment.* Norman: Oklahoma University Press, 1975.

Krier, James E., and Edmund Ursin. *Pollution and Policy.* Berkeley: University of California Press, 1977.

Krier, James E., and Edmund Ursin. Forcing technology: The clean air act experience. *Yale Law Review* ILXXXVIII (1979): 1713–1734.

LASL (Los Alamos Scientific Laboratory). Four Corners Regional Study. Report to NCAQ, Washington, D.C., 1980.

Landau, Jack. Who owns the air? The emission offset concept and its application. *Environmental Law* IX (1979): 575–600.

Latimer, Douglas A. Power plant impacts on visibility in the West: Sitings and emissions control implications. In APCA Specialty Conference Proceedings on Visibility, Denver, Colorado, 1979.

Lave, Lester B., and Eugene P. Seskin. *Air Pollution and Human Health.* Baltimore: Johns Hopkins University Press, 1977.

Liroff, Richard A. Air Pollution Offsets: Trading, Selling and Banking. Washington, D.C.: Conservation Foundation, 1979.

Mansfield, Edwin, et al. *The Production and Application of New Industrial Technology.* New York: Norton, 1977.

Mathtech, Inc. An Analysis Alternative of Policies for Attaining and Maintaining a Short-term NO_2 Standard. Prepared for Council on Environmental Quality, Princeton, N.J., 1978.

NAS (National Academy of Sciences) Environmental Studies Board. On Prevention of Significant Deterioration of Air Quality. Washington, D.C., 1981.

NCAQ (National Commission on Air Quality). To Breathe Clean Air. Report to National Commission on Air Quality, Washington, D.C., 1981.

Navarro, Peter. The 1977 Clean Air Act Amendments: Energy, environmental, economic, and distributional impacts. *Public Policy* 29 (spring 1981): 121–146.

Orr, Lloyd. Incentives for innovation as the basis for an effluent charge strategy. *American Economic Review* LXVI (May 1966): 442–449.

Peltzman, S., and T. N. Tideman. Local vs. national pollution control: Note. *American Economic Review* LXII(1972): 959–963.

Portney, Paul, and Harrison, David. Making Ready for the Clean Air Act. *Regulation* (March/April 1981): 24–31.

Repetto, Robert. The economics of visibility protection: On a clear day you can see a policy. *Natural Resources Journal* XXI (April 1981).

Repetto, Robert. The Influence of Standards, Effluent Charges, and Other Regulatory Approaches on Innovation in Abatement Technology. Center for Population Studies, Harvard University, 1980.

Roberts, Marc, and Michael Spence. Effluent charges and licenses under uncertainty. *Journal of Public Economics* V (1976): 193–208.

Rowe, Robert D., Ralph C. d'Arge, and David S. Brookshire. An experiment on the economic value of visibility. *Journal of Environmental Economics and Management* VII: (March 1980): 1–19.

SAI (Systems Applications, Inc.). An Estimation of the Accuracy and Adequacy of Air Quality Models and Monitoring Data for Use in Assessing the Impact of EPA Significant Deterioration Regulations on Energy Developments. Prepared for American Petroleum Institute, San Rafael, California, 1975.

Schultze, Charles. *The Public Use of Private Interest.* Washington, D.C.: Brookings Institution, 1977.

Sellars, Carolyn A., and William T. Lorenz. Air Pollution Control Industry Outlook: 1979 Update. Boston: Wm. T. Lorenz Co., 1979.

Speciner, Trudy. Getting More for Less: Regulatory Reform at EPA. Office of Policy and Management, U.S. Environmental Protection Agency, Washington, D.C., 1980.

State of California, South Coast Air Quality Management District. Sulfur Dioxide/Sulfate Control Study: Main Text, 1978.

Stern, Arthur C. Prevention of significant deterioration: a critical review. *JAPCA* 27 (1977): 440–453.

Stewart, Richard. Mechanisms of Environmental Regulatory Control and Decisionmaking and Their Relation to Innovation: The Present System and Potential Alternatives. Report to the U.S. Congress by Office of Technology Assessment, Harvard Law School, 1980.

Turner, D. Bruce. Atmospheric dispersion modeling: A critical review. *JAPCA* (1979): 502–519.

U.S. DOE (Department of Energy). National Energy Plan II. Appendix: "Environmental Trends and Impacts." Washington, D.C., 1979.

U.S. DOI (Department of the Interior), Bureau of Land Management. Description of the Outer Continental Shelf Leasing System. Washington, D.C., no date.

U.S. EPA (Environmental Protection Agency), Office of Air Quality Planning and Standards. Emission Density Zoning. 1977.

U.S. EPA. Guideline on Air Quality Models, report EPA-450/2-78-027, 1978.

U.S. EPA. Protecting Visibility: An EPA Report to Congress. 1979a.

U.S. EPA. Methods Development for Assessing Air Pollution Control Benefits, vol. II. Washington, D.C., 1979b.

U.S. EPA. Research Guidelines for Regional Modelings of Fine Particulates, Acid Deposition and Visibility. Washington, D.C., 1980.

U.S. EPA, Office of Planning and Management. The Bubble Clearinghouse. Washington, D.C., 1980.

U.S. EPA. Air Quality Criteria for Particulate Matter and Sulfur Oxides (External Review Draft), vol. II, chap. 2, 1981a.

U.S. EPA. Air Quality Criteria for Particulate Matter and Sulfur Oxides, vols. II–V, 1981b.

Vickrey, William. Auctions, markets and economic theory. In Y. Amihud (ed.), *Biddings and Auctionings for Procurement and Allocation.* New York University Press, 1976.

Walker, Richard, and Michael Storper. Erosion of of the Clean Air Act of 1970. *Boston College Environmental Affairs Law Review* VII (1978): 278–297.

Weitzman, Martin L. Prices vs. quantities. *Review of Economic Studies* XLI (October 1974): 477–491.

Wittman, Donald. First come, first served: An economic analysis of "Coming to the Nuisance." *Journal of Legal Studies* III (June 1980): 557–568.

Yandle, Bruce. The emerging market in air pollution rights. *Regulation* (July-August, 1978): 21–29.

Chapter 14

Air Pollution Control and Solid Waste Recycling, hearings before Subcommittee on Public Health and Welfare, Interstate and Foreign Commerce Committee (March 20, 1970).

Allison, G. *The Essence of Decision.* Boston: Little, Brown, 1971. *Congressional Record* (Senate), November 2, 1971.

Kelman, S. J. *What Price Incentives?: Economists and the Environment.* Boston: Auburn House, 1981.

Kneese, A. V., and C. L. Schultze, *Pollution, Prices and Public Policy.* Washington, D. C.: Brookings Institution, 1975.

Schultze, C. L. *The Public Use of Private Interest.* Brookings Institution, 1977.

Status of the Programs and Policies of the Envirnomental Protection Agency, hearing before Subcommittee on Environmental Pollution, Committee on Public Works, U.S. Senate (95th Congress, first session), January 18, 1977.